GETTING AMERICA READY FOR JAPANESE SCIENCE AND TECHNOLOGY

Edited by

Ronald A. Morse, *Secretary*
Asia Program, The Wilson Center

Richard J. Samuels, *Director*
MIT-Japan Science and Technology Program

THE ASIA PROGRAM OF THE WILSON CENTER
THE MIT-JAPAN SCIENCE AND TECHNOLOGY PROGRAM

University Press of America,® Inc.

4720 Boston Way
Lanham, MD 20706

3 Henrietta Street
London WC2E 8LU England

Printed in the United States of America

Library of Congress Cataloging in Publication Data

Getting America ready for Japanese science and technology.

Preface in Japanese.
Papers from a conference sponsored by the Asia Program of the Woodrow Wilson International Center for Scholars and the MIT—Japan Science and Technology Program.
Includes bibliographies.
1. Science and state—United States—Congresses.
2. Technology and state—United States—Congresses.
3. Scientists—Training of—United States—Congresses.
4. Engineers—Training of—United States—Congresses.
5. Japanese language—Technical language—Congresses
I. Morse, Ronald A. II. Samuels, Richard J.
III. Woodrow Wilson International Center for Scholars.
Asia Program. IV. MIT-Japan Science and Technology Program.
Q127.U6G38 1986 338.97306 86-9089
ISBN 0-8191-5373-7 (alk. paper)
ISBN 0-8191-5374-5 (pbk. : alk. paper)

Because Federal funding was used to prepare this report, "The U.S. Government has a royalty-free, irrevocable, worldwide, nonexclusive license to reproduce, perform, translate and otherwise use, and to authorize others to use, these materials for Government purposes."

ASIA PROGRAM
WOODROW WILSON INTERNATIONAL CENTER FOR SCHOLARS
Smithsonian Institution Building
Washington, D.C. 20560
202/357-1937

DONORS

American Electronics Association

Chemical Abstracts Service

Eastman Kodak Company

The Ford Foundation

International Business Machines Corporation

Mitsui and Company (U.S.A.) Inc.

Morgan Guaranty Trust Company of New York

Motorola Inc.

National Bureau of Standards

National Science Foundation

National Technical Information Service

Nichimen America, Inc.

Nissan Motor Corporation (U.S.A.)

Office of Naval Research

The Procter & Gamble Company

Teradyne Inc.

Tokyo Electric Power Company, Inc.

U.S. Department of Defense

Westinghouse Electric Corporation

Xerox/Japan Technical Information Service

PREFACE

Intent on staying abreast of developments in American scientific research, the Japanese for the last three decades have trained large numbers of translators, assembled technical data bases from American sources, and invested in U.S. training for their scientists and engineers. Yet the United States knows precious little about its chief high-tech competitor. Its researchers are without the language skills to unlock Japanese technical literature and research facilities, and only a handful of American engineering students have ever studied in Japanese universities. The United States needs people with practical (applied) Japanese language skills.

While the U.S. government seems to be uninterested in taking the kinds of actions that would remedy this situation, and the American business community, with a few exceptions, has not known how to fully utilize the skills of Americans with Japanese language and area expertise, a small group of universities and a few concerned groups have attempted to address the issues and to generate debate about the national needs. This report contains their views.

The Asia Program of the Wilson Center and the MIT-Japan Science and Technology Program teamed up to sponsor this conference because of a shared concern that America must be internationally minded and must better understand its major high-tech competitor, Japan, if it wishes to be truly competitive and make the bilateral scientific relationship fully balanced. Scientists, trade association representatives, government administrators, language specialists, and nearly a dozen technical experts from Japan had the chance to sit down together and to examine in a serious way common interests and capabilities. This volume contains the papers presented at the conference and, in the introduction, some lessons drawn from the discussions.

Many people and institutions helped make the conference and this report possible. In particular, we wish to express our thanks to the individuals who travelled from Japan to participate. Takehiko Yamamoto and his team from Bravice International went to considerable expense to give a live demonstration of their machine translation equipment at the conference. Leslie Navarrete, Program Assistant, the Asia Program, made all of the arrangements for the conference. Lisa Olson helped prepare summaries of the papers, Kazuko White prepared the Japanese language summary, and Helen Loerke of the Wilson Center edited and prepared the final manuscript.

The separate list of donors indicates the breadth of interest that is developing concerning this subject. Without their financial support the conference would have been impossible. To the many other people who have assisted in this project, we express our sincere appreciation.

Richard J. Samuels
Associate Professor and
Program Director
MIT-Japan Science and
Technology Program
Room E53-447 MIT
Cambridge, MA 02139
617/253-2449

Ronald A. Morse
Program Secretary
Asia Program
Woodrow Wilson International
Center for Scholars
1000 Jefferson Drive, S.W.
Washington, D.C. 20560
202/357-1937

日米科学技術協力へのアメリカの課題

はじめに

日本は、ここ三十年間、アメリカの科学技術研究に比肩すべく、多大な努力を重ねてきた。数多くの翻訳家の養成、アメリカからの技術データベースの収集、科学者や技術者のアメリカでの研究の奨励など、積極的な方策を立ててきた。

それに対してアメリカは、その主要たるハイテク競争者についてはほとんど何も知らずにいる。アメリカの研究者たちは、日本の技術文献や研究施設を利用できる実用的な（科学技術に関する）語学力を持たず、また、日本の大学や公的研究所で学んだことのある科学・技術者は、ほんの一握りに過ぎない。アメリカが日本の科学技術の発展を十分に理解するために、日本語堪能な技術者や翻訳家を養成することは、何よりも優先されるべきことであるにもかかわらず、全く努力がなされていないのが現状である。アメリカ政府はこの問題に関して無関心であり、議会の公聴会でその必要性が強調されているにもかかわらず、その改善を急ぐ政策、予算に関する法案は実際化されていない。最近になってやっと、アメリカ実業界には、日本語のできる研究者、専門家を採用する会社がでてきたが、これらのごく少数の例外を除いて大多数は有能な人材を十分に活用しているとは言えない。

少数ながら、いくつかの大学や関連団体は、早くから、産業競争力の強化には外国語の理解が必要であることを認識し、これを問題提起して、本格的に検討する機会を望んでいた。彼等の真しにして、熱した声は、この度、本書にまとめられることになった。

ウィルソン国際センターでの学術会議

ウィルソン国際センターのアジア研究科とマサチューセッツ工科大学日本科学技術科は、科学技術に関する日米相互関係の均衡を保つためには、アメリカは日本の科学技術研究にもっと目を向け、理解を深めることが必要であることを強調して、この学術会議を共催することとなった。会議では、科学者、貿易関係代表、政府の担当行政官、現場で日本語教育に携わっている専門家、そして１０人余りの日本からの技術者、翻訳事業関係者が一同に会して、現状の問題点、現状の改善、この問題の解決への可能性を熱心に討議した。本書は、この会議で発表された論文に基づいて構成されている。

アメリカのこれからの重要課題

ーーー　日本語の語学力を持った科学・技術者を養成する重要性を国家的に認識する。

ーーー　実用的日本語教育のために、アメリカ政府そして私的機関は、十分な予算を確保する。

ーーー　アメリカの企業は、日本語の語学力、日本に関する知識を持ったアメリカ人を広く活用する。

そして、この会議で、これらの課題は、緊急に、且つ、慎重に解決されるべきであることが確認された。

CONTENTS

CONTRIBUTORS

HERMAN BARON has been involved in computer-based information systems for the last 15 years. His positions have included employment as an auditor for the Defense Department, a management analyst for a municipal utility, a commercial finance officer, a university librarian, and various managerial positions in the information industry. For the last eight years, he has been an information consultant. He was recently appointed director of Xerox/Japanese Technical Information Service.

R. BYRON BIRD is Vilas Research Professor and John D. MacArthur Professor in the Department of Chemical Engineering at the University of Wisconsin. He received his B.S. in chemical engineering at the University of Illinois, and his Ph.D. in physical chemistry from the University of Wisconsin. He has been a Fulbright Professor at the Technische Hogeschool in Delft, Holland (which awarded him an honorary doctorate in 1977), and at Kyoto and Nagoya Universities in Japan. He has co-authored several books: *Molecular Theory of Gases and Liquids*, *Een Goed Begin: A Contemporary Dutch Reader* (with Shetter), *Reading Dutch* (with Shetter), and *Comprehending Technical Japanese* (with Daub and Inoue). Professor Bird is a member of the National Academy of Engineering and the American Academy of Arts and Sciences.

JUSTIN L. BLOOM, President of Technology International, Inc., is a graduate of the California Institute of Technology with a degree in applied chemistry. He joined the Department of State as director of the Office of Nuclear and Energy Technology Affairs in 1973, and later was Counselor for Scientific and Technological Affairs at the U.S. embassies in Japan and in the United Kingdom.

LARRY D. BROUHARD is president of InterCultural Communications Service (ICCS) in Tokyo, Japan. He formed the company in 1975 to provide a way to gather information for a Master's thesis in sociology at the International Division of Sophia University. ICCS now employs 30 technical translators and rewriters, most of them American, Australian, and English. In addition to work at ICCS, he also teaches seminars for the Japan Society for Technical Communications, is president of the International Computer Association in Tokyo, and oversees the development of computer-aided language-training software for Bydal Systems, Tokyo.

SAMUEL K. COLEMAN received his Ph.D. in anthropology and Japan studies from Columbia University in 1978. His writings include his book, Family Planning in Japanese Society, published by Princeton University Press in 1983. He joined the North Carolina Japan Center at North Carolina State University in 1982 as associate director for Research and Program Development. His current work focuses on academic programs in science and engineering.

EDWARD E. DAUB began his career in chemical engineering, receiving his B.S. and M.S. from the University of Wisconsin, but changed direction in 1947 to enter the ministry. Ordained in the Presbyterian Church, he served two five-year terms as a fraternal worker to the United Church of Christ in Japan, one term doing youth work in churches in the industrial areas of Osaka, the other term teaching chemical engineering and doing student ministry at Doshisha University in Kyoto. Since returning to the States, he received his Ph.D. in the history of science from the University of Wisconsin, taught the history of physical sciences and technology at Kansas University (1965-1971), and presently teaches humanities courses for engineers at the University of Wisconsin.

J. DAVID EDWARDS is currently director of the Joint National Committee for Languages, and the registered lobbyist for the Council for Languages and Other International Studies. Dr. Edwards is also an immediate past chairman of the Council of Washington Representatives, a charter member of the American League of Lobbyists, and a trustee of the International Development Council. Prior to joining JNCL, he was a consultant to the Office of Personnel Management, training senior federal executives in legislation and congressional affairs. A former program director for one of the largest education associations, he first came to Washington as a research associate for a State Department/Department of Defense "think tank" focusing on Latin America. Dr. Edwards has taught at Loyola University, West Virginia Tech, and the University of Virginia.

TATSUYA HAYASHI graduated in applied physics from Waseda University in 1960. In the same year, he joined Fujitsu Ltd., working on the development of basic software. Since then, he has been engaged in software engineering, natural language processing, and AI technologies. He is currently the manager of the software laboratory. He received a prize for the best paper in the field of software engineering in 1973, and has also co-authored several books on computer science.

PAT HILL HUBBARD is vice president of the American Electronics Association (AEA), and founder and president of AEA's Education Foundation. She developed and managed the Association's 1984 pilot program in which six U.S. engineering graduate students received Japanese language training and were sent to Japan to work in research labs for up to a year. Ms. Hubbard holds an M.A. degree from Stanford University and is a doctoral candidate at the University of San Francisco.

RANDALL L. JONES is a professor of German and the director of the Humanities Research Center at Brigham Young University. He earned his Ph.D. in Linguistics at Princeton University, and taught eight years at Cornell before going to BYU. He has been actively using the computer for research and instruction in foreign languages and linguistics for more than 20 years. He is the director of the German TICCIT project, has been involved with several other foreign language CAI projects, and is currently working with CALICO (Computer-Assisted Language Instruction Consortium) on four interactive videodisc projects under contract with the federal government.

ELEANOR H. JORDEN, Mary Donlon Alger Professor of Linguistics at Cornell University, supervises Cornell's Japanese language program, including Japanese FALCON, a full-time, full-year intensive program established in 1972. She is a graduate of Bryn Mawr, and received her Ph.D. in Linguistics from Yale University. A former president of the Association for Asian Studies and of the Association of Teachers of Japanese, she is currently completing a new Japanese language textbook to replace her _Beginning Japanese_, which has been the most widely used Japanese text in the world.

TOSHIHIKO KANDA graduated from Yananashi University in 1962 and joined the Tokyo Electric Power Company. In 1970, he joined JICST, where he developed JOIS (JICST On-line Information System). He is now responsible for marketing coordination, international marketing, and JOIS sales promotion.

TADASHI KORETSUNÉ is founder-president of the National Translation Institute of Science and Technology of Japan (NATIST), and has devoted his career mainly to assisting Japanese in technical writing and in a variety of science and technology related translation projects. Through NATIST's affiliation with F.I.T. (Federation Internationale des Traducteurs [International Federation of Translators]), UNESCO NGO-A, he has worked to protect translators' rights. He has made a major contribution to the professionalization of technical services, particularly through his training institute and its program to certify technical translators. Mr. Koretsuné is a graduate of the Law Department of Tokyo University.

CLAIRE J. KRAMSCH studied at the University of Paris, the Sorbonne, and at the *Ecole Normale Superieure*. She received an *Agregation* in German language and literature. Further studies included work in Munich and at Harvard in applied linguistics. She is currently senior lecturer in the foreign languages and literatures section of the Department of Humanities at MIT. Her writings include *Discourse Analysis and Second Language Teaching*, *Interaction et Discours dans la Classe de Langue*, and a communication strategies textbook, *Reden, Mitreden, Dazwischenreden*. She is one of the principal investigators of the Athena/Annenberg/CPB project for Language Learning through Interactive Computer Software.

RICHARD D. LAMBERT is professor of sociology of South Asia at the University of Pennsylvania. He was president of the Association of Asian Studies (1974-75), and is editor of *The Annals*, the Journal of the American Academy of Political Science. An expert on international studies, he is also the editor of *Beyond Growth: The Next Stage in Language and Area Studies*, published by the Association of American Universities in 1984.

RONALD A. MORSE is Secretary of the Asia Program, The Wilson Center. He has a Ph.D. from Princeton University. Before joining The Wilson Center in 1981, he worked at the Department of Defense and at the State Department. He is an editor and contributor to the book, *The Politics of Japan's Energy Strategy*, and edited a study on artificial intelligence in preparation for U.S. participation in Expo '85 in Tsukuba, Japan. In 1983 he was a member of a National Academy of Sciences mission to Japan on high-technology ceramics.

JANET H. MURRAY is project director of the Athena Language Learning Project funded by the Annenberg/CPB Project, and of the Athena Writing Project. She holds a Ph.D. in English literature from Harvard University, where her work included study in Latin, French, Spanish, and German. She worked as a systems programmer for IBM in 1967, and has more recently been studying software engineering at MIT. She has taught at MIT since 1971, and since 1980 has been affiliated with the Experimental Study Group, which specializes in individualized instruction. She is the author of numerous articles and a popular textbook, *Strong-Minded Women and Other Lost Voices from Victorian Literature* (Pantheon, 1982; Penguin, 1984), as well as radio scripts for National Public Radio. She is currently working on a book on the development of the computer as a medium for language and literature.

ROBERT A. RUSSELL is currently an associate professor of Japanese, and coordinator of the M.A. program in Language Acquisition at Brigham Young University. He received his Ph.D. in Linguistics at Harvard University, and taught Japanese for four years at the University of Hawaii. Before going to BYU, he spent two and a half years in research and development, serving as the director of linguistics of a machine translation firm. He has published articles and read papers on various topics of machine translation, linguistics and language acquisition, and computer-assisted language instruction, and is currently the principal investigator on a Japanese computer-assisted instruction project funded by the U.S. Department of Education. He is director of the Japanese TICCIT/CAI project at BYU. He serves on the board of directors of the Association of Teachers of Japanese, and is active in a number of other professional organizations.

RICHARD J. SAMUELS is director of the MIT-Japan Science and Technology Program and Mitsui Career Development Associate Professor of Contemporary Technology in the Department of Political Science. His book, _The Politics of Regional Policy in Japan: Localities Incorporated_, was published in 1983 by Princeton University Press. His most recent research has been on public energy corporations in industrial democracies, particularly Japan and Italy. He has a Ph.D. from the Massachusetts Institute of Technology.

TOMOYUKI SATOH has operated Japan Information Service, a prestigious Japanese-English technical translation company in Austin, Texas, for the past three years. His clients have included many of the Fortune 500 corporations, research associations, and top researchers in the United States. He attended the Science University of Tokyo and holds an M.A. in physics from the Johns Hopkins University. He has also done graduate work in architecture. He was recently appointed associate director of Xerox/Japanese Technical Information Service.

J. MARSHALL UNGER is associate professor in the Department of East Asian Languages and Literatures at the University of Hawaii at Manoa. He has a Ph.D. in Linguistics from Yale (1975), and his dissertation was on the earliest recorded forms of the Japanese language. He was a Japan Foundation fellow during 1985, researching the connection between the Fifth Generation project and the Japanese script-handling problem. His most recent publications are "Japanese Orthography in the Computer Age" and "Japanese Braille," both in _Visible Language_ 18:3 (Summer 1984).

GALAL WALKER received his Ph.D. in East Asian literature from Cornell University. He has taught Chinese with the Cornell FALCON Program and at the University of Pennsylvania. At present he is teaching Chinese language and literature at Ohio State University. Current projects in Chinese language pedagogy include the development of individualized instruction courses in Chinese and organizing the Chinese School for the East Asian Summer Language Institute at Indiana University.

TAKEHIKO YAMAMOTO studied law at the University of Tokyo. He founded Bravice International Inc. in 1971, and in 1982 became the chairman of the board of WCC, Chicago. He is a member of the Japanese National Automatic Translation Telephone Development Committee.

INTRODUCTION

The United States faces a number of critical issues with Japan, and the problem of America's technical illiteracy about Japanese science and technology is a cause for national concern. Whether the United States can train sufficient numbers of engineers to be language proficient and school enough translators to adequately understand Japanese scientific and technological developments should be (but is not) a top national priority. As one conference participant pointed out, the present U.S. attitude on this subject is economically, technically, and intellectually unwise.

Policymakers have remained indifferent to a number of studies on this subject. Executive Branch assessments of critical national language priorities have been largely ignored. The United States-Japan Advisory Commission to the President of the United States and the Prime Minister of Japan (September 1984) stated, "A variety of measures, including increased Japanese language training for Americans, should be taken to enlarge opportunities for scientists from both countries to work together and improve access to each other's research." To date there has been no follow-up. The U.S. government's lack of in-house Japanese expertise has been evident for some time.* Congressional hearings have shown a dire need for both language training and programs to make access to foreign science and technology easier, but follow-up and funding bills to make these needs a reality have gone nowhere.**

---

* "Japan's Bureaucratic Edge," by Ronald A. Morse and Edward A. Olsen, Foreign Policy, no. 52, Fall 1983, pp. 167-80.

** See The Availability of Japanese Scientific and Technical Information in the United States (Hearings before the Subcommittee on Science, Research and Technology of the Committee on Science and Technology), U.S. House of Representatives, March 6, 7, 1984; and the Report on the same topic prepared by the Congressional Research Service, Library of Congress, and transmitted to the Committee on Science and Technology, U.S. House of Representatives, November 1984.

Only recently has American business shown an interest in hiring Japanese-speaking researchers or area specialists trained in the United States. The January 1985 report by the President's Commission on Industrial Competitiveness points to the lack of "critical information about foreign markets" and the need to "strengthen our ability to compete," but never acknowledges the need for reading and understanding foreign languages. A National Academy of Sciences report (1984) on High-Technology Ceramics in Japan recommended the development of a "vitally needed mechanism for gathering and disseminating timely information on Japanese ceramics publications, reports and patents," but the corporate sponsors of the report have remained indifferent to the idea. In 1984, the American Chamber of Commerce in Japan for the first time established a research and development subcommittee to get a handle on monitoring the progress of Japanese technology.* Information from Japan collected by U.S. public agencies, while voluminous, is not made available to industrial or academic users.

Applied (practical) Japanese studies will not develop easily. Unless something is done, young Americans will be reluctant to invest their time and effort in language study if it does not demonstrably enhance their career opportunities. Moreover, scientists trying to learn Japanese have found text materials and course schedules inappropriate. Others are hard-pressed to get started because qualified Japanese language teachers are so scarce. It is clear that there are no shortcuts to learning particularly a high-culture-context language like Japanese.

Traditional Asian language and area-study centers at universities that have received considerable federal funding have just begun to adjust to new demands for Japanese instruction for scientists and engineers. None has redirected major resources toward technical students. MIT offers science and engineering students a one-year internship in Japanese public and private laboratories, but it is leading an uphill battle.** And only four other U.S. universities -- North Carolina State, Wisconsin, New York University, and Lehigh -- have start-up programs in applied Japanese studies.

---

* "America Starts Looking Over Japan's Shoulder: U.S. Industry and Government Belatedly Boost Their Monitoring of Japanese Technology," Business Week, February 13, 1984.

** R.B. Gillmor and R.J. Samuels (eds.), Japanese Scientific and Technical Information in the United States (Proceedings of a Workshop at MIT), January 1983. Reproduced by National Technical Information Service, U.S. Department of Commerce, Springfield, VA 22161.

Should U.S. demand for Japanese science and technology ever awaken, new language technologies in the form of computer-assisted instruction, including videodiscs and computer-student "dialogues," may be useful. A translating machine, demonstrated at the conference by Bravice International President Takehiko Yamamoto, was impressive in "translating" a Japanese sentence into good English. As with so many other areas, the Japanese now dominate the machine translation market. For now, however, the conventional commercial translation companies that abstract, translate, and perform a wide range of on-line data services seem to be the only answer to accessing Japanese technical information. But this is no substitute for the benefits that can derive from researchers working in Japanese facilities and participating in Japanese scientific meetings.

While the United States has moved from apathy to concern on this problem, it has not yet turned the corner about what to do. Part of the problem is that there is still no focal point for leadership in the business community or in government. This conference moved the national agenda forward and helped to identify the issues that need to be more systematically addressed. But as everyone recognized, there is now a critical need for follow-up and action in the national interest.

ANYONE READING THIS REPORT WILL RECOGNIZE WHAT IS NEEDED

-- a new business/government awareness of the importance of training scientists and engineers in the Japanese language;

-- increased federal and private sector funding for applied Japanese language studies; and

-- a commitment by U.S. corporations to employ Americans who acquire foreign language and area expertise.

I

JAPANESE TRAINING FOR A NEW GENERATION OF SCIENTISTS AND ENGINEERS

ARE JAPANESE LANGUAGE PROGRAMS REACHING SCIENTISTS AND ENGINEERS?

Eleanor H. Jorden

A memorable theatrical review appeared a number of years ago that read something like this: "At the Middletown Repertory Theater last night, James Morrison played Hamlet. Hamlet lost."

Today we are speaking not of acting a part, but we are speaking of reaching an audience -- and obviously the quality of our performance has a direct bearing on our success. While there have been recent signs of a dramatic improvement over the past, if an ultimate tragedy is to be averted, we must act promptly to produce more meaningful results.

Recently a review of the language programs in one of our major universities was brought to my attention. It was particularly interesting for two reasons: one was the fact that Japanese had been ranked as clearly the best program in the university; and second was the fact that the program rated the worst was described as having "no faculty with training or academic interest in language teaching." In a general statement included in the report, it was pointed out that "language teaching takes place in departments that are basically literature departments...and where the prestige lies with literature." I include these latter comments in spite of the fact that they were not written about Japanese, but because they could apply to many Japanese programs elsewhere. The fact remains that basically our language teaching heritage in the United States derives from the European tradition, which regularly focussed on literature as the academic pursuit that accounted for language as a subject of serious study. Ignore the fact that in actual practice the so-called reading of great works of literature in language classrooms has usually involved no more than the painful decoding of passages, accompanied by more dictionary than literary reading, and the blackening of the page with interlinear glosses to be used during those excruciating classroom translation sessions. With the reading of literature the goal, from the first day of instruction everything points in that direction.

On the continent of Europe, geography assists in the addition of oral skills, and a general assumption of a multilingual world helps to extend students beyond an exclusively literary focus. But the literature heritage of language teaching, imported into the United

States, has had little leavening, and it persists significantly in many quarters, including some of our most prestigious colleges and universities. If you are seeking out the language program at a college campus in this country, you are most apt to find it physically and administratively joined with literature, with literature studies identified as the intellectual pursuit of the department.

Those Japanese high school students who are made to struggle with Shakespeare in their English classes while still at the most elementary proficiency level are Asian victims of a similar heritage. Hamlet loses again. As a Japanese high school principal once proudly told me, "There isn't an English sentence our students can't puzzle out." How tragic an approach for those who study language for communication rather than puzzle-solving.

Lest you think my comments are anti-literary, let me explain that I am only anti-literature-as-the-only-reason-for-studying-language; and and that I am also anti-identification-of-literature-teaching-and-language teaching, for the goals, analysis, and teaching techniques -- to say nothing of the required teacher training -- are distinct. Of significance to us today is the fact that this literary tradition has undoubtedly been a factor in the development of a general assumption that scientists and engineers don't study foreign languages beyond satisfying a language requirement, and for that purpose one looks for the "easiest language" -- rarely identified as Japanese. But times are changing. People travel these days, to a degree previously undreamed of, and when abroad, an inability to communicate proves extremely frustrating. It is particularly infuriating to those individuals who spent long hours studying the language in question in academic courses.

I like to think of attitudes toward the acquisition of foreign language proficiency in terms of three commonly occurring, basic types. Type A can be defined as the "If-we-need-to-communicate-let-them-learn-English Type." This was the unenlightened attitude that described a shockingly high percentage of the American public until very recently. And it included a particularly high percentage of American-born scientists and engineers who had contact with foreign counterparts.

But as a nation, we are now transferring into Type B -- which I describe as the "Looking-for-a-quick-fix Type." Americans are always interested in prompt and rapid solutions to problems, and foreign language acquisition is no exception. The Japanese, who are always careful to accommodate the market place, have produced many answers to the American desire for a quick fix in the texts they produce: "Japanese in 3 Weeks," "Japanese in 30 Hours," "Japanese for Busy People," "Speak Japanese Today," etc. And new methodologies that claim miraculous results -- proficiency in no time at all with no effort at all -- are further examples. Just last week I received a telephone call from the secretary of a businessman interested in a one week immersion course to prepare himself for Japan. My apologetic reply -- that our introducductory full-time, intensive course lasted a calendar year -- must have been not only a disappointment but also a shock. Another request a few years ago, for a "2-hour per week course over 3 or 4 months, with limited goals -- just to enable the senior executives of a major corporation to understand Japanese non-technical conversation" -- was one more example of Type B.

But we must never underestimate the progress reflected by the move from Type A to B. Herein lies the realization that English alone is not enough, that study of foreign languages and cultures is necessary for international interaction, and that in the case of Japan most particularly, serious problems have resulted from our persistent Type A attitude.

Increasingly scientists and engineers are among those moving into Type B. From there progress to a Type C attitude, which acknowledges the need for systematic study of a foreign language in depth, may perhaps be only a matter of time. The quick-fix approach is usually of short duration, in that it quickly proves its own inadequacy; and while its casualties may drop back to a Type A attitude, they will at least be <u>wiser</u> Type A's, who <u>know</u> that dependence on English is a reflection of inability or unwillingness to devote the required time and effort to foreign language acquisition. In many cases, this knowledge is now accompanied by regret and an awareness that a Type A attitude leads to less than optimal international interaction and communication. But outweighing the casualties will be the successes -- those who move on quickly to a Type C attitude and continue in their efforts long enough to reach significant and substantive foreign language competence. The question for us to consider today is whether the scientists and engineers of Type C are being adequately nurtured. Are they able to continue their study of Japanese to the point where they can function linguistically in their area of specialization?

In 1983, an in-depth survey of language and area programs in the United States was made at the request of Congress. The Association of American Universities organized the study, with funding from the National Endowment for the Humanities and the Department of Defense.

The final report was <u>Beyond Growth</u>, prepared by Richard Lambert. In my capacity as a member of the investigating team concentrating on language programs, I had a rare opportunity to view a cross-section of programs in the uncommonly taught languages, including Japanese. My reactions ranged from considerable optimism to rather deep depression.

Generally speaking, Japanese is in a much more favorable position than the average program in the commonly taught languages. To a considerable degree, the uncommonly taught group received their first significant thrust from the intensive programs of World War II, which, for all their weaknesses, did recognize that foreign languages could actually be used for oral as well as written communication. In addition, academics trained in linguistic analysis were presented with the challenge of analyzing these languages and preparing teaching materials, which exerted some influence on many of the texts subsequently produced. And Japanese -- like the others in this category --

has unquestionably gained from the fact that it is regularly studied by students who want to study it: there are indeed very few students of Japanese who have chosen it as the least demanding way to satisfy a language requirement.

Thus, we might find that on many campuses, Japanese would rank among the more successful local language programs -- even as the best. But "best" -- depending on the competition -- may cover a range from superb and without peer anywhere, to least bad among local evils. I believe it was one of our vice presidents whose reply to the question "How's your wife?" was "Compared to what?" I hasten to point out that in the report cited above, the description of the Japanese program made clear that it was indeed "most good" rather than "least bad."

Leaving aside for a moment the Japanese language programs that are being offered with professional expertise at all levels -- and there definitely are such programs -- there is a traditional pattern of organization that is still encountered all too frequently. Advanced Japanese classes and seminars -- with low enrollments -- tend to be offered by the most prestigious scholars and are usually literary in focus. At some institutions, there are also individual specialists in history, philosophy, religion, etc. who take directed reading courses, often on a tutorial basis, with one of the native-speaker instructors. As we move down in the scale of courses toward the more elementary, there is an inverse relationship between the number of students and the experience of supervisory professors: the enrollments at the introductory level are enormous, handled all too often by the newest professor on the staff. The theory seems to be that the rider who has never been on a horse is best suited for the horse that never has been ridden.

If the professors were only new, the situation would not be as serious. But all too often, their academic training has focussed on literature -- literary criticism or comparative literature or literary translation. Depending on the actual individuals, they will regard their assignment to introductory language classes at best as a fascinating and stimulating new challenge but, in at least some cases, as a period of purgatory before full-time admission to the heaven of advanced literature seminars. In both instances, almost without exception they are untrained for the assignment in foreign language pedagogy. Even with the best of intentions and the best possible results, they will find themselves re-inventing the wheel after time-consuming trial and error.

Lest I seem to be criticizing only literature specialists, I would point out that theoretical linguists -- with no additional pedagogy-related background -- are no better off. What is needed is relevant, specialized training in the field of pedagogical linguistics, and our recognition of this field and its importance has been shamefully lacking. Perhaps in no other academic area have we permitted non-specialists such a loud voice -- even to the point of making language policy.

Might an answer to the requirement for skilled instructors be native speakers of the target language? Not without specialized training! Yet, how many language teachers are there today, in program after program, in country after country, teaching their native languages to unknowing and uncritical foreigners. I wince whenever I overhear conversations on our campuses that run something like this: "I hear you want to go to Japan. There are lots of interesting jobs over there, but if worst comes to worst, you can always teach English." Worst coming to worst, indeed!

The Japanese teaching staffs in institutions across the United States, then, range from truly excellent to poor, with comparatively few trained specifically for the task of teaching Japanese to Americans. Almost without exception, they follow a locally determined curriculum that encompasses elementary, intermediate, and advanced instruction. The definitions of these levels are precisely imprecise: elementary = what we do in first year; intermediate = our second year curriculum; and advanced = our third year. "We" and "our" refer, of course, to the institution where the statements are made. Beyond the third year, small numbers move into tutorials and seminars that are, almost without exception, reading courses.

According to the recent Lambert report, 78 percent of students studying Japanese in the institutions surveyed did not continue beyond second year Japanese; and had that figure been broken down into first year versus second year termination, the rapid attrition rate would undoubtedly be much more dramatic. Enrollments in U.S. college and university Japanese language courses have shot up proportionately faster than those of any other foreign language. The 40 percent increase between 1980 and 1983 recorded by the Modern Language Association is impressive. This increasing population of students includes budding scientists and engineers, but how many will continue to a significant level of competence in Japanese? How many will become an attrition statistic prematurely?

Enrollment attrition may be a reflection of student dissatisfaction. There will always be a number of students who, from the start, plan only one year of Japanese study. However, many are simply turned off by the method and approach. When we hear reports that at some institutions the only continuing students in a second year course were those with Japanese parents or those who had spent a significant period of time in Japan, we suspect a first-year curriculum that is unrealistic for the true beginner.

Recently I received a letter from a former student, now a professor of Japanese. "Enrollments are at record levels," she wrote, "although a fourth of the students...have not returned for the second quarter. Part of the loss, I'm sure, was due to the less than ideal conditions: class only four days a week, sections set at 35 students, an antiquated language lab that is not open in the evenings and will not allow students to check tapes out overnight, a set-up that requires the native speaker and me each to take full responsibility for two sections instead of team teaching."

Another crucial factor must not be overlooked. By the time students have finished one or two years of instruction, they are acutely aware of the long-term commitment required to achieve professional competence in Japanese, perhaps longer than they care to commit themselves to. French was never like this! But to some extent we are involved in a Catch-22 situation: Scientists and engineers often have no incentive for continuing into advanced Japanese because there are no specialized programs for them; since they don't continue, there is no reason to set up special programs.

For those who do want to continue, it is unusual to find any departure from a single-track language curriculum in any institution. And even if administrators could be persuaded to experiment with advanced courses specifically designed for special purposes, who would teach them? The problems of recruitment of staff for such courses are significant, particularly when the "special purpose" involves science or engineering. Few specialists in these fields, with competence in Japanese in addition, are interested in becoming language instructors.

We do have a critical need for scientists and engineers with a strong Japanese language capability. I would describe this need not simply in terms of ability to read the literature, crucial though that may be, but also in terms of ability to communicate directly with Japanese counterparts. If reading were the only requirement, why would so many Japanese be in the United States actually observing our industries and interacting with American specialists? This need must be emphasized at the highest administrative levels of American business. The Japanese have long realized the importance of sending personnel to the United States for on-site participation in American projects, but this has been a one-way street. Americans, traditionally locked in Type A attitudes, have been unable to interact with non-English speaking Japanese, of whom there are many in Japan.

Now we are waking up at last. Last summer the American Electronics Association mounted a program for engineers that entailed an intensive summer language program with heavy emphasis on socio-linguistic orientation, followed by an internship in a Japanese corporation in Japan. The program has proved so successful that it will be continued and doubled this year. Originally, the AEA had planned to recruit engineers who already had a Japanese facility. They discovered very quickly that up to that point, the graduate engineer and the Japanese-proficient American represented a mutually-exclusive grouping. The fallback position was to recruit the engineers and start them out in the language.

But this is only a beginning. For countless Americans, to move to what is now called the "novice-mid" level of proficiency in Japanese -- even the "novice-low" level -- will be a great stride forward, but we also desperately need true specialists. To quote the late Henry Lee Smith, "Nine women each pregnant for one month will never produce a baby." We need some Americans who can make an appointment with a Japanese counterpart on the telephone, find their way to his lab, introduce themselves, make appropriate small talk, and then discuss research of mutual interest covering discussion of relevant publications -- all in Japanese and using not only appropriate vocabulary and structure, but appropriate politeness levels and terms of address and timing.

This represents a great commitment of time and effort. Although graduate engineers just starting out in Japanese will probably never achieve this level of proficiency, they should be encouraged to progress as time permits to receive the tremendous benefits realized by even the lower levels of competence. However, we must also recruit at the lower levels, urging students to begin the study of Japanese while they are still in an earlier stage of their professional training.

If there were more courses for special purposes, might there not be less enrollment attrition? For the scientists and engineers who continue Japanese language studies, there are few offerings of courses relevant to their needs. Courses for business purposes are starting to be offered in some institutions, but advanced language curricula specifically for scientists and engineers will require significant additional pedagogically-oriented effort. My own assessment of the need is for the kind of solution I suggested in a paper delivered at the Association for Asian Studies in 1980. The trained pedagogical linguist, drawing on the topical expertise of members of the science and engineering faculties, could develop the kind of linguistically sophisticated, pedagogically productive and "engineeringly accurate" materials that would be truly purposeful. The kind of approach Galal Walker will be discussing might be adaptable for the preparation and utilization of such materials. With a minimum of additional funding and staff, they could be incorporated into the Japanese curriculum of many institutions, but I must emphasize the necessity for professionalism in the original preparation of the materials, a formidable task.

It is at the point where the linguistic and structural basics are no longer a mystery that the student would be ready to move into specialized language -- which after all should be primarily a vocabulary problem. But the teaching of these advanced levels must not be identified with the knowledge of them. How and by whom they are to be taught must be given as much attention as the language material itself. An all-too-common attempt to find a quick solution to our dearth of language specialists is noted in the rush into specialized reading prematurely. The eventual problems are enormous -- aside from the sheer drudgery that accompanies any attempt to cope with foreign language material that is beyond one's level of competence. In the end, what appears to be increased speed may develop into a slowdown -- and even the attrition of potentially high-caliber specialists.

Years ago, I was visited frequently by beginning students who needed "just the vocabulary for Zen Buddhism" or some such highly specialized technical field. Today there seems to be slightly more sophistication about language study.

Research into machine translation is important in that it promises ultimately to make possible the production of comprehensible and accurate equivalents of Japanese technical articles. It should be given high priority in order to make possible the widespread dissemination of information on Japanese research among the entire scientific community. But this will not satisfy the additional requirement for true Japanese specialists among our scientists and engineers. Nor will the requirement be met by American scientists and engineers who are able only to decode articles in their field. Even technical articles can be read with a different kind of understanding when one has worked on the entire framework of the Japanese language, and sees where this scientific portion fits into the whole. In other words, I would urge starting off engineers the same way all students should begin -- with a solid introduction that shows how this language is put together and how it reflects the society in which it is used, and how the Japanese reflect their high-context culture in every facet of their existence.

At no point in the foreseeable future will every American college and university with a Japanese language program be able to offer programs in Japanese specifically for scientists and engineers taught by specialists. But alternate approaches for reaching this group should be established. We need further research and development in the areas of individualized instruction, self-instruction, and computer-assisted instruction, always keeping in mind that only the subject matter relates to special purposes: how it is to be taught is a question of foreign language pedagogy.

A high priority recommendation of the Lambert report involved centers for serious research in language pedagogy. One of the topics that cries for attention is special purpose language training -- particularly the development and utilization of course materials. A science curriculum suitable for widespread use should reflect the most advanced level of foreign language pedagogy as well as good science. Given the degree of specialization in the modern world, we had better separate out areas of science. The effort must be cooperative. Hopefully the product will be not decoders who live with their dictionaries but communicators who not only read their Japanese counterparts' output but also meet and interact with them.

Are Japanese language programs reaching scientists and engineers? No -- but scientists and engineers are beginning to reach Japanese language programs! And if they are to reach a professional level of Japanese language competence in their respective fields, our programs must reach a new level of diversification.

JAPANESE LANGUAGE STUDY FOR ENGINEERS: HIGH TECH PERSPECTIVES

Pat Hill Hubbard

A few weeks ago a billboard ad appeared alongside a Silicon Valley highway. Five or six new Dodge Colt automobiles are shown lined up beneath a simple slogan: "Colt: All the Japanese you ever need to know."

I was struck by the irony of this ad. I had just finished reading Iacocca's editorial calling on the federal government to deal with Japan's unfair trade advantages or else, he threatened, American industries will have to "relocate to foreign countries, or go out of business."[1]

On an obvious level, Dodge/Chrysler's ad is paternalistically telling us not to worry about learning Japanese in order to compare the quality of foreign imports. "Trust us, your U.S. distributor. We have made the comparisons for you." Subliminally, however, the ad works to lull this country's growing awareness that perhaps we do need to understand the language of the Japanese, since their presence has become so pervasive in our economic lives.

The Japanese national policy is clearly focussed on scientific and technological advancement. As a result, Japan is outpacing the United States in the worldwide trade race.

U.S. Electronics Industry: Its Nature and Impact

Until recently, the United States held an unchallenged leadership position in world trade, underpinned by its preeminence in high technology.

U.S. based electronics companies now have worldwide sales of over $300 billion and domestic revenues of $185 billion. With an anticipated 15 percent annual revenue growth, electronics sales in the United States alone are expected to reach $500 billion by 1990. While electronics is now the sixth largest U.S. industry sector, by the turn of the century it expects to be second only to energy. Electronics is already the largest single manufacturing segment in the United States, currently employing 2.5 million Americans. In the next 15 years, the computer and peripheral equipment industry will become the fourth fastest growth industry in the country, creating 48.3 percent new jobs.

We are already significant job creators in states such as Massachusetts, California, Florida, Minnesota, metropolitan New York and Texas -- centers of electronics and information technology companies. And for every job created within a high tech company, the White House Cabinet staff tells us another eight jobs are created in other industries that service or supply it. In addition, our high technology industries produce the technological tools that all economic sectors -- banks, automobile manufacturers, steel and oil companies, etc. -- are using to increase their productivity and economic competitiveness.

A 1983 U.S. Department of Commerce study underscores high tech contributions to the U.S. economy. Almost in "House that Jack Built" terms, the high tech economic foundation is described as:

> ...a direct linkage between the research activities conducted by high technology industries and the U.S. standard of living. Research nurtures innovation, which leads to productivity gains...which determines the overall ability of the U.S. economy to grow and in turn to produce a higher standard of living and new jobs.

The report cautions, however, that over the last 12 years, U.S. high tech industries' trade position has declined "from dominance to one of being strongly challenged" and that "market share for the high technology group -- and for nearly all individual industries -- has fallen."[2]

This is an alarming shift. Electronics, along with agriculture, has traditionally been on the positive side of the trade ledger. It has always been other industries that have been part of the erosion from a positive trade surplus of $9 billion in 1975 to a $69 billion deficit in 1983. While only four years ago, the U.S. electronics industries enjoyed a $6 billion surplus, in 1984 we had a deficit of $6.8 billion. For the first time, we are part of an unprecedented $123 billion worldwide U.S. trade deficit -- triple what it was in 1982.

Japan, on the other hand, ended last year with a $31 billion surplus from the United States alone. Japan's Research Institute on the National Economy projects the country's overall surplus over the next four years will total a whopping $255 billion.[3]

When U.S.-Japan trade in electronics-based products is spotlighted, the major source of our growing imbalance blazes forth. The U.S. electronics deficit with Japan grew from almost $4 billion in 1980 to over $15 billion last year (see Attachment A). It is expected to be $20 billion by the end of 1985.

Japan is the free world's Far East cornerstone, a strong ally in any threat to world peace. Yet in the trade arena, as one Asian expert recently put it, "The technological battle with the Japanese is really the industrial equivalent to the East-West arms race."[4]

Japan's Trade Advantages

Japan's "trade army" is well equipped: 99 percent literacy, a strong work ethic and company loyalty all generate high productivity. Labor costs in Japan are 50 percent of ours, and the cost of capital for electronics is two to three times less than in the United States. Other key trade advantages are:

Ready Availability of Capital: Of the major industrialized nations Japan has the highest rate of personal savings and capital formation. Unlike the U.S., where high tech companies generally rely on public stock offerings and venture capital, Japan's equity financing comes from banks. Of the 30 largest banks in the world, in fact, 11 are Japanese.[5]

Government Policy Reinforces Economic Goals: Japan's national policies have long included economic goals. The country's consensus approach to decision-making creates an environment in which industry interacts with rather than reacts to government, as in the United States. Interaction creates a strategy which targets and focusses national resources on winners. William Tanaka, in 1981 testimony before the Joint Subcommittee on Trade, Productivity, and Economic Growth, described it:

> They [the Japanese] systematically allow their "sunset" industries to expire if they cannot compete. Protection in Japan is accorded to "sunrise," knowledge- and information-oriented industries....And even once-favored industries are allowed to fail as the technological process of innovation moves on.[6]

Japanese protection of its sunrise industries from marketplace intrusions by foreign competitors logically follows.

Strong Commitment to Engineering Education: Since Independence in 1951, Japan's emphasis has been on strengthening education's role to achieve economic success:

> ...economic competition among nations is technical competition, and technical competition has become an educational competition.[7]

Engineering education has been emphasized. In 1982, with only one-half the U.S. population, Japan produced 10 percent more engineering graduates with bachelor degrees (BSEs). Japan surpassed us in BS graduates in 1967 and has continued to do so ever since.

Excluding foreign nationals from U.S. degree data, Japan now graduates over twice as many BSEs as the United States.[8] We do produce more students at the engineering masters and doctorate degree levels. However, without foreign nationals in the U.S degree pool, and combining all three degree levels, we still have significant shortfalls (see Attachment B).

Access to U.S. Technology: As the technological world leader, we have successfully promulgated the idea that English is the international language of research. Japanese require six years of English language study prior to secondary school (high school) graduation, and an additional 12 credits in college.[9] "This major linguistic advancement over American counterparts" allows "them to read the literature of at least the English-speaking countries." Even in the University of Tokyo library, less than half the books and publications are in Japanese or Chinese.[10]

As alumni and investors in U.S. university-based research, the Japanese are pouring money into our engineering schools. They have become a major market for licensing U.S. university-developed technology. Through university affiliate programs and endowed chairs, they get early access to technological developments. U.S. universities, they seem to believe, are "simply the best place to train future researchers for Japan."[11]

It is important to keep our universities free and open to all students. We benefit when foreign graduates return home. Trained in our technology, they become employees of our foreign subsidiaries, and frequently initiate trade licensing agreements with us.

Almost all of Japan's engineering students are citizens. Yet in 1982-83, 1,097 Japanese were engineering students studying in the United States; of these, 318 were electrical engineering majors. Only a dribble of American engineering students have ever studied in Japanese universities. Larry Grayson, of the National Institute for Education, notes: "In the last 20 years, not more than seven Americans were enrolled in Japanese engineering programs in any one year, and for most years there were none."[12]

Japan Moves to Technological Leadership Role: Traditionally Japan has emphasized quality and reliability in its products. It has looked to other nations, especially the United States, for products to duplicate or to improve. Today, Japan is changing roles. It is moving from a technological follower to a technological leader.

Even by 1981, Japanese R&D spending was $26 billion -- some four times that of the previous decade. On a per capita basis, this is equal to the United States' $52 billion civilian R&D investment. Moreover, the Japanese government supports 25 percent of R&D, with only 2 percent spent for defense. The U.S. government, however, while it supports twice as much R&D, spends 70 percent for defense.[13]

R&D investment is fostering Japanese trade success. An American Chamber study in Tokyo found that U.S. technology was being "overtaken in 12 important areas, including advanced ceramics, optical fibers, and large scale integrated circuits."[14] And since a decade ago, the Japanese have been selling to us more technology than we sell to them. They hold 40 percent of all patents worldwide -- four times the number of U.S. filings.[15] The National Superspeed Computer Project and the Fifth-Generation "Thinking" Computer Systems Project are well-known efforts by Japan to take the technological lead.

"Through a Glass Darkly": U.S. Scientific Myopia

In spite of the view of some that technical information exchange at the governmental level between the United States and Japan is basically "steady state,"[16] growing evidence points to the contrary in terms of industrial information.

Dr. James R. Bartholomew of Ohio State University's History Department, in testimony last March before the House Subcommittee on Science, Research, and Technology, maintains:

> It's not correct that the Japanese are putting most of their stuff in English....Most engineering research effort in Japan is done in the private sector, specifically the private corporation sector....Information produced is of a proprietary nature; it's hard for us to gain access to that.[17]

Company research is reported privately through in-house publications and is, therefore, unlikely to become part of a country-wide data-base that U.S. translators can access. Dr. Bartholomew states that only 25 percent of Japan's some 10,000 technical journals and literature now appear in English.[18] Further, current translation costs, ranging from $.10 to $.25 a word,[19] make price exorbitant. And adding to expense is a dearth of language experts with the technical competence to translate technical material when it is accessible. This makes U.S. sight of Japanese technological breakthroughs almost predestined to be dim in comparison to the constancy of the clear 20-20 of our Japanese trading partners.

How AEA is Responding

The American Electronics Association is the largest U.S. trade association for electronics and information technology companies. Nationwide, our 2,700 member companies contain all segments of the industry. AEA members account for 63 percent of the $300 billion in worldwide sales of U.S. based electronics industries. They include manufacturers and suppliers of computers, semiconductors, components and peripherals, telecommunications equipment, defense systems and products, office systems, instruments, software, and R&D firms. Our membership ranges from the largest companies to the start-ups; 80 percent are small, employing fewer than 250 people.

Our members have been concerned for some time about the threat to U.S. technological leadership posed by Japan's commercial inroads. AEA is responding on many fronts -- in Washington and in Japan.

The U.S. presence in Japan is growing. Of the slightly more than 200 U.S. companies with operations in Japan, 126, or over 60 percent, are members of AEA. Of those, we have employment data on 145, which indicates they employ collectively 12,000 people. With funding assistance from the U.S. Department of Commerce, AEA opened a trade office in Tokyo last summer to assist U.S. firms to garner a more equitable share of the Japanese electronics market. Among other services, AEA/Tokyo will be a resource for technical information to assist companies to design and manufacture competitive products for the Japanese marketplace.

Last February, in anticipation of opening this trade office and in cooperation with Japanese private industry and the Electronics Industries Association of Japan, AEA, under its Electronics Education Foundation, began the AEA Japan Fellowship Program. The program's goal is to foster development of a mutual appreciation of the technological capabilities of both countries. The approach is to increase the number of U.S. electrical and computer engineering and computer science graduate students who have knowledge and sensitivity to Japanese language and culture and firsthand experience in Japanese company research laboratories.

Our program design is to send U.S. graduate students to Japan for up to a year to do research in company labs. We wrestled immediately with whether or not they needed to have Japanese language capability. Our Japanese industry friends assured us it was not necessary, since English is spoken in their laboratories. Some of our AEA company executives were of the same mind. Yet we did not intend these students to live like expatriates. They would live like Japanese workers -- housed in company dorms and fed at company cafeterias. Also, we recognized that the informal network of relationships they would build with their Japanese co-workers off the job would form the long-term technical communication links for the future. While we did not see it necessary for them to be fluent in reading or writing or to be able to translate, we stood fast in our idea they should at least have some speaking and reading facility. Some Japanese language study would help them while they were in Japan, and would as well establish a base on which to continue coursework at a later date.

Unlike a single university that generally must draw from matriculated students, AEA as a national organization was free to draw from many universities. We initially thought that from a graduate student universe of some 28,000 students in our targeted fields, we would find some Issei or Nissei with native-speaking ability, or appropriate technical students who had taken Japanese courses. We were naive! We searched through Japan Societies and at universities with well-reputed Asian language programs. We found at the universities either no graduate engineering departments or no technical students who could speak Japanese and who met our criteria.

The most salient reasons we found for engineering students -- especially at the graduate level -- to avoid Japanese programs on their campuses were:

(1) Japanese language classes are geared to language majors, held one hour a day four days a week, etc., and conflict with research and class schedules of engineering majors.

(2) Japanese class content is geared to art, history, and literature and has little relevance to the interests of technical students.

(3) The heavy academic and research demands on engineering graduate students for upwards of two years for a masters and five to six for a doctorate do not allow space for any serious language study.

We decided, therefore, to find an intensive summer language program to which we could send our students as a group, thus offering them also the opportunity of supplementing language study with Saturday seminars on cultural and technical topics of interest. We selected the program at Cornell University, and with Eleanor Jorden and Jeff Frey's help, our six students all passed the course. Jeff, acting as AEA's Japan onsite coordinator, accompanied five of the students to Japan last fall. Two have already returned, three are still there, and one is going next month.

The feedback on the usefulness of their language study is unanimously positive. One student, Doug Browning, states it well:

> As I suspected would be the case, the language program has proven to be invaluable. It is perhaps possible to survive here without knowing Japanese, so many people at the office speak English well and almost all read it; but when I think of the lost oportunities for interaction with the Japanese people both in the office and elsewhere from knowing no Japanese it makes me cringe.

We intend to double the Fellowship program to 12 students this fall. We hope our recipients will continue language study and that, over time, we will build a cadre of technical people capable of enhancing our electronics trade position with the Japanese.

Other High Tech Perspectives on Language

It is important to point out that, while most member companies are enthusiastic about our fellowship program -- some 13 have supported the language study with hard dollars -- mixed perspectives exist on the usefulness of Japanese for technical employees.

We completed last week a survey to quantify the language needs of some of our members doing business in Japan. Thirty-four or 49 percent of those companies sent questionnaires responded. Some highlights of the responses:

Types of Business Operations in Japan

76% sales and service
35% engineering
26% manufacturing
15% regional headquarters
8% R&D or Procurement

Number of U.S. Citizens at Japan Operations

Of the 6,517 combined numbers of workers at the 34 reporting companies, each has on average only one to two U.S. citizens in company employee populations that average 97 workers.

Need for Japanese Language Training

40% reported never a need
47% reported sometimes or infrequently a need
13% reported often a need
1 company reported a need in the future

Level of Need

Of those 60 percent who see a need for language training, 89 percent see minimal to fluent speaking capability as necessary. The next highest need is for reading, expressed by 63 percent. A need for technical vocabulary was reported by 58 percent.

While 58 percent see no need to learn to write and 63 percent no need to translate, 16 percent feel it necessary to be fluent in writing and 26 percent to be fluent in translation capability.

Problems in Getting Japanese Language Training

Problems respondents identified in order of those most frequently mentioned:

(1) Finding competent instructors
(2) Expense
(3) Time-consuming/impact on work schedules
(4) Convincing domestic managers of the worth in time and cost
(5) Inconveniently scheduled local classes

No Japanese language preparation is provided by 55 percent of the companies for their workers in Japan. Of those who do provide it, 18 percent study with private language training institutes (most common is Berlitz), 12 percent offer employees courses in Japan with private tutors or through local language schools.

Fifty-three employees from 23 companies took some form of Japanese language training in 1984.

Additional comments

In keeping with a traditional worldwide perspective on staffing foreign subsidiaries, we hire locals to conduct our business. We hire only Japanese in Japan.

It is more important to know how the Japanese do business, how to negotiate with them, Japanese pricing policies, and cultural aspects of doing business in Japan.

The biggest problem we face is getting English training for Japanese employees.

We could use Japanese training for U.S. employees who interact with Japanese as well as for travel in Japan.

One company executive in a personal interview gave extensive comments (see Attachment C).

No other electronics trade association or research groups currently have activities to equip technical professionals with Japanese language training.

The Austin-based MCC anticipates in the future that twenty of their technical researchers will need training in speaking and reading Japanese. They currently employ two U.S. citizens to translate technical journals and articles. They contract with outside translators as well, but find it time consuming, expensive, and hard to locate individuals with technical backgrounds. They state that only 5 to 10 percent of all Japanese technical journals are currently translated.

Steps To Be Taken

The Department of Commerce has already taken the first steps to help us develop a model through partial financial support of our Japan coordinator. We have also proposed to the National Science Foundation that they provide a "challenge grant" to AEA's Foundation to match, dollar for dollar, monies raised from private industry for our fellowship program's summer language component.

Beyond our program needs, however, we need to develop a national consensus that Japanese language training for engineers and technical people is important. Out of that public policy position may come many creative ideas. One, suggested by an AEA member, is to establish a federal Center for Japanese Science and Technology. Similar to the training for diplomats offered by the Foreign Service Institute, this new Center could translate technical journals, could offer Japanese language study geared to engineering and technical people, and could provide funds and/or technical assistance to selected university language departments to design and offer special classes for engineering students.

The need is growing. The federal role is less one of dollars than of leadership.

FOOTNOTES

1. Iacocca, Lee A., "Wanted: A Game Plan for America," <u>Spectator</u>, Detroit, Michigan, Winter 1985, p. 5.

2. <u>U.S. Competitiveness in High Technology Industries</u>, U.S. Dept. of Commerce, International Trade Administration, Washington, D.C., February 1983, p. iii.

3. "Trade Talks with Japan May Not Head Off Collision with Japan," <u>Business Week</u>, December 31, 1984, p. 39.

4. Gresser, Julian, President East-Asian Consulting Group, "Japan's High Tech Challenge," <u>Newsweek</u>, August 9, 1982, p. 48; Grayson, Larry P., "Japan's Intellectual Challenge: The Strategy," <u>Engineering Education</u>, Washington, D.C., December 1983, p. 139.

5. Grayson, "...The Strategy."

6. Tanaka, H. William, Statement before Congressional Joint Economic Committee's Subcommittee on Trade, Productivity, and Economic Growth, June 23-28, 1981, Washington, D.C., p. 86.

7. Shinahara, K., Adaptation and Education in Japan (New York: Praeger) 1979, pp. 129-130; Grayson, Larry P., "Japan's Intellectual Challenge: The Future," Engineering Education, Washington, D.C., February 1984, p. 140.

8. Grayson, "...The Strategy," p. 142.

9. Ibid., p. 139.

10. Grayson, Larry P., "Japan's Intellectual Challenge: The System," Engineering Education, January 1984, p. 218.

11. "Japan is Buying Its Way into U.S. University Labs," Business Week, September 29, 1984 (Research), p. 178.

12. Grayson, "...The Strategy," p. 145.

13. Ibid., p. 142.

14. "America Starts Looking over Japan's Shoulder," Business Week, February 13, 1984 (Research).

15. Bartholomew, James R., Hearings before House Subcommittee on Science, Research, and Technology, March 6-7, 1984, p. 90.

16. Bloom, Justin L., Hearings before House Subcommittee on Science, Research, and Technology, March 6-7, 1984, pp. 3-5.

17. Bartholomew, p. 91.

18. "It's Time U.S. Started Picking Japan's Brains," Christian Science Monitor, August 12, 1984.

19. Dee, Paula, "The High Cost of Monitoring Japanese Technology," Electronic Business, October 1, 1984, pp. 86-87.

ATTACHMENT A

Electronics Trade Worldwide

Year	Imports	Exports	Balance of Trade
1980	$15.5 billion	$21.8 billion	+ $6.3 billion*
1981	$19.2 billion	$24 billion	+ $4.8 billion*
1982	$21.4 billion	$24.9 billion	+ $3.5 billion*
1983	$26.7 billion	$26.8 billion	+ $100 million**
1984	n/a	n/a	- $7 billion**

Source: * AEA
** U.S. Dept. of Commerce

U.S. Trade with Japan in Electronics-Based Products 1980-84 (In Millions of Current Dollars)

	1980	1981	1982	1983	1984
Consumer Products					
Exports	62	50	37	33	35
Imports	2,337	3,646	3,410	3,985	6,040
Deficit	-$2,275	-$3,596	-$3,373	-$3,952	-$6,005
Industrial Electronics					
Exports	1,562	1,802	1,904	2,154	2,639
Imports	3,136	4,240	5,109	7,197	12,082
Deficit	-$1,574	-$2,438	-$3,205	-$5,043	-$9,443
Total Products					
Deficit	-$3,849	-$6,034	-$6,578	-$8,995	-$15,498

Source: U.S. Dept. of Commerce
International Trade Administration
Trade Development, Science, and
Electronics Sector

ATTACHMENT B

Comparison of Engineering Degrees - U.S. and Japan

Engineering Degrees Produced 1982:

Degree Level	Japan*	U.S. Less Foreign Students' Degrees	Difference
Bachelor	73,593	61,580	-12,013
Masters	7,363	13,073	+ 5,710
Doctorate	621	1,720	+ 1,009
			5,204**

Electrical Engineering Degrees Produced 1982:

Degree Level	Japan*	U.S. Less Foreign Students' Degrees	Difference
Bachelor	19,429	14,758	- 4,671
Masters	1,671	3,089	+ 1,418
Doctorate	132	229	+ 188
			3,065***

* Japan engineering students are almost all citizens.

** U.S. produced 5,204 fewer non-foreign engineering students at all degree levels combined than Japan.

*** U.S. produced 3,065 fewer non-foreign electrical engineering students at all degree levels combined than Japan.

ATTACHMENT C

Comments from Personal Interview with George Scalise, Senior Vice President and Chief Administration Officer, Advanced Micro Devices, on Need for Japanese Language Training for Engineers and Technical People, Santa Clara, California, January 4, 1984:

The U.S. Government has concluded that it is not important we know the Japanese language. This is the reverse from the Japanese, who believe it extremely important that their scientists and engineers learn English. The United States has become convinced that English is the international language. This has been fostered by many countries and by specific examples such as the International Transportation System, which requires that all pilots and controllers learn English. English's use internationally is borne out in so many ways that the United States population has become lazy about over-adopting this position.

Such attitudes are detrimental. If our chief competitor is Japan, and if Japanese can readily understand and participate in all U.S. technical symposia, but the U.S. cannot participate in any of the Japanese ones, this gives the Japanese a decided trade advantage.

The solution is to encourage broader educational programs for engineering students. Some form of foreign language -- hopefully Japanese -- for two years or more should be a requirement for engineering students. High schools, which now commonly teach French, Spanish, and German, rarely teach Japanese. Post-secondary education, therefore, becomes the college of last resort to learn the language. This is extremely important for engineers, although we also need plant managers and financial people who can speak and understand Japanese. As a company in Japan grows, it is even feasible that the personnel person would be a U.S. citizen.

Basically, an engineering graduate who has "special skill" in understanding and speaking Japanese is eminently more hireable than one without.

Companies are accommodating to the need. One way is by having key people in Japanese subsidiaries be understudied by other bilingual technical people. This avoids management problems related to language in the case of employee turnover. If a company gets caught where it needs to replace someone in Japan but has no one who knows Japanese, they would likely approach the problem by taking a technically competent person and teaching them language rather than taking one who speaks it and try to instill them with technical capabilities. While there is a need to raise the consciousness of the companies, the universities, and the students to the need to increase Japanese language training for engineers and other technical people, the role of the federal government should be to lead in raising the consciousness and using federal dollars to seed start-up efforts. It might, for example, focus dollars to develop teachers with the capability to teach Japanese language courses. It might, through NSF, fund pilot projects to teach engineering students the language -- such as the AEA Japan Fellowship Program model.

APPLIED JAPANESE STUDIES FOR SCIENCE AND ENGINEERING AT AMERICAN UNIVERSITIES

Samuel K. Coleman and Richard J. Samuels

Recent scientific and technical developments have been punctuated by a new appreciation of the importance of Japan as a source of knowledge and the potential commercial application of this information. The American business and technical community has begun to seek ways to identify and to utilize Japanese research. Studies and conferences sponsored by government,[1] universities,[2] and the private sector[3] have confirmed the uniformly high quality of Japanese research and have recognized the need for timely access to it by Americans. These studies and meetings have also confirmed the problems and difficulties associated with identification and dissemination of the results of Japanese research. The dearth of technologically sophisticated Americans with research experience in Japan and with Japanese language capabilities is most often cited as the critical bottleneck.

That such persons have not been produced in sufficient numbers is, in our view, not a Japanese but an American problem. Its solution is properly the responsibility of our universities -- with public and private support. The purpose of this paper is to assess the state of such "Applied Japanese Studies" in our institutions of higher education, with particular reference to the activities of designated centers of Japanese studies.[4] We shall identify how programs might be expanded to better serve national goals.

Although a broad consensus has developed that access to Japanese scientific and technical information requires some number of American scientists and engineers to attain a professional command of the Japanese language,[5] these specialists who would be educated about Japan have needs that go beyond what is ordinarily satisfied by technical curricula on the one hand or by traditional Asia studies curricula on the other. In the case of the former, they need information and analyses of the broader context in which Japanese research and development take place. Understanding the social, political, and economic organization of Japanese research is properly a part of a student's education, not a matter left for on-the-job training. Much can be accomplished through coursework to orient the future scientist and engineer toward the relationships among R&D, industrial policy goals, technology transfer, and international competitiveness that obtain in Japan. Likewise, much can be done to equip the student with effective knowledge about interpersonal relations, invisible colleges, and the dynamics of Japanese-style research collaboration and decision-making.

In the case of Asian studies curricula, education about Japan must be effectively balanced with the rigorous demands of technical education. In the process, our traditional criteria for Japan expertise might well have to be reassessed. In order to begin helping to create many more technically sophisticated graduates with the ability to operate professionally in Japan and in Japanese, we should accept the possibility that our efforts will result in a new breed of Japan specialists who look very much unlike ourselves.

Solutions to these challenges may take several forms. Language teachers, confronted by a new cohort of students with non-traditional motivations for studying Japanese, will have to respond with new materials, some of which utilize scientific and technical themes. Interestingly, their new students will possibly be more comfortable with the new technologies of language learning than the teachers are themselves. Japanologists will also find themselves interacting directly and continually with science and engineering faculty -- often for the first time. Their collaboration will serve to help articulate alternative career opportunities for science and engineering majors who attain proficiency in Japanese.

Study programs abroad should be broadened to encourage technical students, and internship opportunities should be broadened to include placements in Japanese laboratories and industries. Hands-on experience is a sine qua non for any graduate of a science or engineering program who claims Japan-related expertise. Only through the research enterprise itself can the student fully comprehend the differences in American and Japanese research strategies and styles, and acquire the knowledge to become an effective seeker and user of Japanese scientific and technical information. Only in Japan can the student absorb the myriad details of Japanese life that influence -- and are influenced by -- that country's science and technology. This sort of early research experience in Japanese universities and industry also affords an opportunity to create contacts that will serve the student well for the rest of his/her career.

Innovative language programs and expanded research placements are but part of what programs in applied Japanese studies might offer. The programs that most effectively meet students' needs will be those that also serve the Japan-related professional needs of scientists and engineers in the academic community, reinforce the efforts of the Japanese language and area specialists, and articulate the benefits of the program to business and government. Faculty members in the sciences and in engineering who do not themselves take up the study of Japanese can nonetheless make use of Japanese scientific and technical information through library/data base support either in English or in Japanese, translated by experts in the Japanese study centers. The more these faculty benefit from enhancement of their own network of Japanese collaborators, the more enthusiastic they will become in supporting their own students' exploration of Japan. By promoting on-campus facilities and resources for access to Japanese scientific and technical information and by helping to screen and certify non-traditional students of Japan, Japanese language teachers become invaluable colleagues to both the technical faculty as well as to area specialists. The considerable benefits of these new relationships, both within and beyond the campus, are easily imagined.

Current Developments in Applied Japanese Studies for Scientists and Engineers

The authors queried faculty representatives at each of the 15 HEA National Resource Centers for East Asian International Studies[6] and also consulted colleagues at other universities who direct programs in applied Japanese studies for scientists and engineers. Our questions concerned three areas of academic activity: Japanese language training, student support and curriculum development, and information dissemination.

The first set of questions sought to measure the participation of science and engineering students in Japanese language courses and to identify offerings in technical Japanese. The second set of questions was designed to identify internships and scholarships provided by these centers for technical students, as well as course offerings covering Japanese science, technology, and society. We also asked about the number of science and engineering students in joint degree and certification programs in Asian studies. In our final category, we asked questions about activities to bring information about Japanese scientific and technical issues to science and engineering faculty, the business community, and to other specialists. The responses to our queries are recorded in Figure 1.

Figure 1 - Applied Japanese Studies at Title VI Centers

	LANGUAGE		STUDENT SUPPORT				INFORMATION DISSEMINATION		
	Percent S/E Majors Among Students Enrolled in Japanese	Technical Japanese Language Courses	Internships	Scholarships	Joint Majors (Current or Average Yearly Enrollment)	Courses on Japanese Science, Technology, and Society	Outreach	Library Holdings in Japanese S&T	Japanese S&T Research Activities
Center/Question	1	2	3	4	5	6	7	8	9
Berkeley	n/a	N	N	N	0	N	Y	Y	Y
Columbia	n/a	N	N	N	0	N	Y	N	N
Cornell	11	N	Y	N	0	Y	Y	Y	Y
Harvard	n/a	N	N	N	0	Y	Y	Y	Y
Hawaii	6	N	N	N	0	N	N	N	Y
Illinois	21	N	N	N	n/a	N	N	N	Y
Indiana	1	N	N	N	3	N	N	N	N
Kansas	1	N	N	N	2	N	N	N	N
Michigan	7	N	N	N	n/a	N	Y	Y	Y
Ohio State	21	N	Y*	Y*	0	Y	Y	Y	Y
Pittsburgh	5	N	N	N	2	N	N	N	N
Princeton	26	N	N	N	1	N	Y	N	Y
Stanford	17	N	N	N	0	Y	Y	N	Y
Washington	7	N	N	N	0	N	N	N	Y
Yale	10	N	N	N	n/a	N	N	Y	N

NOTE: These data are imprecise. They reflect the most recent information available, but are occasionally non-comparable due to institutional differences in definition, and difficulties in their collection. For example, the proportion of technical students we report enrolled in Japanese language classes at Cornell University reflects only those in the Engineering School, and students in the physical sciences there are excluded. The asterisk refers to programs planned but not yet in operation.

In addition to the activities reported above, we can also report on those of non-designated centers, including our own:

LEHIGH UNIVERSITY: In 1979 Lehigh University began an East Asian studies program geared to the University's strengths in science, technology and business. Under the program, eight engineering students have already graduated with a minor in East Asian studies. The program will place three students in IBM internships in Japan for periods of six months to one year. The internship currently has no Japanese language requirement. The university has received a two-year grant for 1982-84 from the U.S. Department of Education, and additional support, including a grant from the Japan Foundation, to enrich library resources in Japanese studies. Outreach activities include a recent conference for regional business people and state government officials on the subject of Japanese science and technology.

THE MASSACHUSETTS INSTITUTE OF TECHNOLOGY: MIT hosts 150-200 visiting researchers from Japan each year. In addition, some 50 Japanese firms are members of the MIT Industrial Liaison Program. In recent years, the Institute has expanded its own teaching in Japanese studies by adding faculty with research experience in Japan, both in the Sloan School of Management as well as in the School of Social Sciences and Humanities. In 1981, the MIT-Japan Science and Technology Program was established to promote a balanced dialogue between scientists and engineers in the United States and Japan. The Program has been supported by Japanese and American corporate contributions and by the internal resources of the Institute. There are three principal elements in the Program: education, research, and public affairs.

The first of these activities is most relevant to this conference. Interns who have studied Japan and Japanese, often with fellowship support from the Program, are placed in public, private, and university laboratories in Japan. The Program trains and supervises eight to ten interns per year, all of whom are supported financially by their host institution. Interns have been placed in the central research laboratories of many of Japan's largest private firms, including Matsushita, Hitachi, NEC, and Toshiba. Others, usually two per year, have been admitted to the Engineering Faculty of Tokyo University. A separate agreement between the Faculty of Engineering at Kyoto University and the Department of Civil Engineering at MIT has also resulted in placements of several students.

In the past, students have cross-registered at Wellesley College and at Harvard University to study Japanese. In January 1985, the MIT-Japan Science and Technology Program sponsored an experimental course in intensive Japanese. Although places were limited to 15 students, over 50 applied. This demand resulted in the introduction of a course in the Japanese language at MIT, in collaboration with Wellesley College, in the fall of 1985.

NEW YORK UNIVERSITY: The Japanese Translation Project at NYU originated in 1980 on the basis of recommendations made by the Computer Science and Technology Board of the National Research Council, which recommended that selected Japanese papers in computer science be translated on a continuing basis. One important by-product of that translation effort has been the creation of a Japanese concentration for computer science majors. A three-year sequence of Japanese language classes for computer science majors was formally added to the university curriculum as a pilot program in 1983. Demand from computer science students has been overwhelming, and new sections have had to be added. Nearly 40 percent of all students taking Elementary Japanese at NYU were computer science majors during 1983-84.

NORTH CAROLINA STATE UNIVERSITY: Activities in Japanese science and technology at NCSU are an outgrowth of the North Carolina Japan Center, established on the university's campus in 1980 to promote the state's economic, academic, and cultural relationship with Japan. The Japanese language program has become the largest in the eight-state southeast region. Over the last several years, some two-thirds of the students enrolled have majored in science and engineering. Through the Harry C. Kelly Fund for U.S.-Japan Scientific Cooperation, the Japan Center is structuring an awards program for the most promising students in the sciences and engineering who successfully undertake Japanese language study. This fund also offers support for Japanese scientists and engineers from non-profit institutions for collaborative research.

NCSU has sent two students to IBM-Japan's Fujisawa Development Laboratory, and one has received a Fulbright fellowship for research at the Tokyo Institute of Technology. The Japan Center's program in science and technology also helps facilitate access to Japanese scientific and technical information by providing information and referrals for technical translating services, and by working with the university's library to acquire technical reference books and dictionaries.

UNIVERSITY OF WISCONSIN: The University of Wisconsin is the only major American university currently offering a course on technical Japanese for its engineering students. The course was developed over a decade ago by Professors Daub and Bird of the College of Engineering. With the collaboration of Professor Inoue of the Science University of Tokyo, they co-authored *Comprehending Technical Japanese*, the first textbook on the subject of technical Japanese-English translation. Twelve students enrolled in the Technical Japanese course between 1981-84. The course has been offered through the University's Extension Program, and associated faculty have also made available a statewide technical translation service.

In addition, the UW College of Engineering has recently concluded agreements for cooperation with Kyoto University and with Kobe University for the exchange of students. Accompanying these agreements has been the development of a separate "U.S./Japan Engineering Leadership Program." This five-year undergraduate curriculum combines the standard requirements of a bachelor's degree in engineering with extensive Japanese studies, including one year's residence at a Japanese university and an internship in a research laboratory.

Conclusion

Until recently, there was little reason for government and industry to pay careful attention to Japanese scientific and technical developments. There was therefore little reason for universities to do likewise. This is no longer the case. If American scientists and engineers are to have the same degree of access to Japanese science and technology that they enjoy vis-a-vis Europe, and if our researchers are to be able to engage in genuine scientific and technical cooperation with Japanese colleagues, then our curricula will need to reflect our national needs.

The evidence we have gathered indicates that these needs have been joined by widespread demand for information about Japan by scientists and engineers. Science and engineering majors account for between one-sixth to one-quarter of those enrolled in Japanese language courses in some universities, such as Illinois, Ohio State, Princeton, and Stanford. At others, such as at New York University and North Carolina State, the percentage is higher.

This combination of need and demand, in turn, has engendered new courses and research activities related to scientific and technical issues among Japan specialists. Faculty affiliated with designated centers have begun reaching out to business and across to colleagues in science and engineering. For example, Cornell University has worked with the American Electronics Association to train interns in Japanese, Ohio State University is about to initiate its own Japan Science and Technology Program, and Harvard University has worked with the Massachusetts High Technology Council to find summer employment for students majoring in East Asian studies. More focussed initiatives have been taken by centers of applied research, such as at MIT, North Carolina State, NYU, and the Engineering College at the University of Wisconsin, where scientific and technical subjects are more central to the curriculum than are Japanese studies. Our sense is that we are observing the early stirrings of a broadly based interest in applied Japanese studies at our universities.

But it seems to us that need, demand and supply are imbalanced. Put differently, these develoments seem demand-driven from the bottom up, rather than from the top down. The significant representation of science and engineering students in Japanese language courses is almost entirely the result of student initiative. This initiative is particularly striking given competing curricular demands and the absence of explicit recruitment efforts. There is virtually no internship and fellowship support for science and engineering majors at the leading centers of Japanese studies. There are likewise no courses in technical Japanese. We infer that the dearth of science and engineering students who undertake a dual concentration in Japanese studies is at least partly attributable to this lack of active marketing. While the growing acceptance of applied Japanese studies seems a response to demand from students and industry, it has not yet been accompanied by the kinds of shifts in institutional orientation that are necessary to meet the needs articulated above.

We believe that Japan specialists at American universities can continue to make genuine contributions toward the development of a cohort of sophisticated scientists and engineers educated about and experienced in Japan. The number of these students remains extremely low, however. Our sampling leads us to an informed guess that there are no more than 400 scientific and technical specialists-in-training now receiving Japanese language instruction in the United States. Moreover, this estimate only represents those currently receiving instruction, not those with proficiency in the language. The inadequacy of this level of enrollment is dramatized by the fact that these individuals represent less than two-tenths of one percent of America's annual degree recipients in the sciences and engineering.[7]

We have suggested curricular modifications and have identified efforts to enhance the role of Japan specialists in technical education; this would in turn raise the number of science and engineering students who study Japan. Designated centers of Japanese studies, especially those with affiliated science and engineering schools, can be key resources in this regard.[8] But applied Japanology will also continue to be important at universities with extensive scientific and technical curricula that are not traditional centers for research on Japan. The task is to encourage both groups of universities to approach this issue more comprehensively. On the one hand, traditional centers can begin to market their offerings to technical students, and on the other hand, our leading institutions of science and technology must continue to expand their offerings in Japanese area studies, including language.

FOOTNOTES

1. The JTECH panels sponsored by the U.S. Department of Commerce seem to be the most comprehensive public sector examinations of Japanese technology and industrial competitiveness to date. These assessments cover Micro- and Opto-electronics, Megatronics, Computer Science and Biotechnology. A final report was to be issued in January 1985 by Science Applications, the sub-contractor.

   Another influential report, "Industry to Industry International Armaments Cooperation Phase II - Japan," was prepared in June 1984 by the Defense Science Board Task Force for the Office of the Under Secretary of Defense for Research and Engineering.

   See also Hane, et al., "Assessment of Technical Strengths and Information Flow of Energy Conservation Research in Japan," prepared for the U.S. Department of Energy by the Battelle Memorial Institute (September 1984); Herman W. Lewis, "Biotechnology in Japan," National Science Foundation (June 1984); Richard Dolen, "Japan's Fifth Generation Computer Project," Scientific Bulletin,

Department of the Navy Office of Naval Research Far East (July-September 1982); George E. Lindamood, "The Rise of the Japanese Computer Industry," ibid., (October-December 1982); Carrie J. Hunter, "U.S. Federal Agency Interest in Developments in Japanese Science and Technology," an unpublished Master's Thesis submitted to the Department of Political Science at MIT (February 1985) provides a useful overview of these and other government activities.

2. For university-based assessments of Japanese technology and of the transparency of Japanese research activities, see Reginald B. Gillmor and Richard J. Samuels, eds., "Japanese Scientific and Technical Information in the United States," National Technical Information Service, 1983; George Kenney and Kent H. Bowen, "High Tech Ceramics in Japan: Current and Future Markets," American Ceramic Society Bulletin (May 1983); Samuel K. Coleman, "Japan and North Caroline State University," Occasional Papers of the North Carolina Japan Center, no. 2 (November 1983).

3. There have been a great many private sector assessments of Japanese scientific and technical activities. See for example, the "Second Japan Issue" of the American Ceramic Society Bulletin (September 1984); David Baskerville, "Sources of Electronics Information in Japan," American Electronics Association (March 1985); "High Technology Ceramics in Japan," National Academy of Sciences (December 1984); "International Developments in Computer Science," National Academy of Sciences Report (1982).

4. For the purposes of this conference, we restrict ourselves to "Applied Japanese Studies" in the sciences and engineering. This term also refers to applications in a wide variety of professional fields such as journalism, architecture, law, and management, among others that have received increased attention in the past decade.

5. See the published testimony before the Subcommittee on Science, Research, and Technology, House Committee on Science and Technology, March 1984.

6. These centers, designated by the Higher Education Act of 1981, are listed in Figure 1. Our survey did not include inquiries into separate programs undertaken by schools of science and engineering at these centers. In the course of our conversations with representatives of these centers, we also asked for help in identifying other colleagues involved in the sorts of activities covered in our survey. This "snowball" sampling revealed no programs beyond those included in this report.

7. Based upon figures in National Science Foundation, Science and Engineering Degrees: 1950-1980, Special Report NSF 82-307.

8. The only universities with Title VI Centers that do not have schools of engineering are Harvard and Indiana.

TRAINING AND CERTIFYING JAPANESE TECHNICAL TRANSLATORS

Tadashi Koretsuné

I shall first explain the background of the two functions of our program for certifying and training Japanese Technical Translators at the National Technical Institute of Science and Technology (NATIST).

Over a stretch of some five years back in the early 1960s, while I was in the United States, France, the United Kingdom, and West Germany, I had the experience of reviewing and assessing the quality of a huge number of business documents sent overseas from Japanese industries, as well as a fairly large number of papers sent overseas from Japanese academic circles. The English language in those business documents and academic papers was, in many cases, found to be either inaccurate or inadequate in style and format, while others were found to be incomprehensible. Still others were found to lack the dynamism needed in the presentation of the products and technologies being offered to the potential business partners.

With a few exceptions -- such as Sony Corporation, NEC (Nippon Electric Company), Fujitsu, Panasonic, and several joint-venture companies in Japan of American or German multi-national companies -- a majority of Japanese companies doing business in the United States and/or doing business with American companies, prepare English-language business documents that are below standard as compared with those from, for instance, West Germany.

Despite these findings, I believed that there must be people in industries and research institutions in Japan capable of preparing their business documents and research reports in English acceptable to their foreign counterparts, and that the companies and research organizations had assigned the wrong employees to such responsibilities simply because they lacked an adequate method of selecting those best suited for preparing documents and papers intended for overseas.

However, a series of surveys we conducted with some 550 leading Japanese companies, patent bureaus, and academic circles proved my belief to be too optimistic and not to reflect the true state of affairs. You can see the survey results in Exhibit A.

An inquiry was made by the U.S. Patent Office as to whether the Japanese Patent Office could find some method to improve the English written in the Japanese patent specifications intended for filing in the United States. The U.S. Patent Office found that a substantial number of the Japanese patent specifications were difficult to understand because of unclear descriptions in the English translations. To cope with these situations, NATIST was organized in 1966, and in November 1967, the first language proficiency test was conducted. Since that time, the test has been conducted twice yearly (in May and November); the test was given for the 36th time in November 1984. The test is administered by the National Board of Examiners for Foreign Language Proficiency Test in Science and Technology.

Tables 1, 2 and 3 give some statistics on the tests that have been given:

Table 1

	Accredited Grade					
Specialty	1st	2nd	3rd	4th*	NG	
Mechanical Engineering	2	22	34	42	28	
Electrical Engineering		11	40	35	23	
Chemistry/Pharmaceutical/Medical	1	13	70	47	11	
Metallurgy/Mining		7	9	7	2	
Civil Engineering/Architecture		1	5	7	1	
Agriculture/Fishery/Forestry		2	7	6		
Systems Engineering		6	13	11	1	
Applied Physics/Nuclear Science	1	2	5	6	1	
Industrial Property Rights		8	8	11		
Licensing/Technical Cooperation) Foreign Trade/Legal Documents)		12	69	80	11	
Totals	4	84	260	252	78	678**

* Not a full-fledged translator (4th grade)

** Of the 678 candidates, 623 are specialists in English and the rest are in French, German, Russian, and Spanish

Table 2

Pass and Fail Rate (31st through 36th/1984): 1st, 2nd, and 3rd Grades only:

	31st	32nd	33rd	34th	35th	36th	Total
Candidates	516	508	519	499	678	511	3,231
Pass	202	192	201	177	222	196	1,190
Pass rate	39%	37%	38%	35%	32%	38%	36% (average)

NOTE: Table 2 shows that the average Pass Rate is 36%. However, the breakdown indicates (as seen in Table 1) that 1st Grade represents only 0.58%, or less than 6 for each 1,000 candidates, 2nd Grade 12.39%, and 3rd Grade 38.35%.

Commemorating the "35th," a special drive was undertaken with major Japanese companies and research institutions, urging them to have their employees take the Test. We had more candidates in this particular Test than we had had in previous Tests. Although the overall Pass Rate was low, those sent from industries were accredited to higher grades than non-corporate candidates.

Table 3

The Number of Candidates Who Take the Test More Than Twice (36th Test)

	2	3	4	5	6	7	8	9	10/24	Total
Candidates	131	92	51	28	13	11	8	7	18	359*

* This represents approximately 70% of the total candidates (511), which means a majority of candidates have taken the Test more than twice.

Successful candidates who have been certified as 1st, 2nd, and 3rd Grades are called "Licensed Technical Translators," while those certified as 4th Grade are called "Associate LTT's." Only the full-fledged LTT's are qualified for practising professional translation, and NA-TIST's unpublished regulations provide that only 1st and 2nd Grade LTT's are qualified for two-way translation (English into Japanese and Japanese into English), while 3rd Grade LTT's are qualified only for one-way translation (English into Japanese).

EXHIBIT A

Surveys were conducted on two different occasions through professional survey groups in Japan. The survey of 1965 showed the following results.

A. Research Papers, Technical Reports, Licensing Agreements, Technical Assistance Agreements, and Patent Specifications

1. Partly incomprehensible?	Yes	77.3%	No	13.0%
2. Possible misunderstanding/misinterpretation because of ambiguity or obscurity?	Yes	60.8%	No	29.8%
3. Rewriting/editing by native speakers of English necessary?	Yes	68.4%	No	7.9%

B. Operating Manuals and Product Specifications

1. Partly difficult to understand because of ambiguity or obscurity?	Yes	66.8%	No	20.1%
2. Inconsistence/irregularity in usage of technical terminology or business terms?	Yes	70.2%	No	27.6%
3. Misleading phrases/expressions that tend to cause wrong operation of machines?	Yes	66.2%	No	30.5%

C. Catalogues and Business Guides

1. Substantial improvement considered necessary?	Yes	76.8%	No	15.3%
2. Poor descriptions resulting in degrading company/product image?	Yes	80.2%	No	13.8%

D. Technical Materials and Feasibility Reports in International Bidding

1. Inconsistency in usage of technical terminology?	Yes	62.5%	No	21.0%
2. Sentences could be interpreted in two or more different possible meanings?	Yes	77.3%	No	18.8%
3. Rights and responsibilities/duties ambiguously defined?	Yes	80.3%	No	12.6%

STANDARD OF EVALUATION

Criteria have been established by NATIST as the basis for the evaluation of Foreign Language Proficiency Tests to be held under its Statute (Article 4, Clause 1) and the resulting accreditation of the grades to be awarded to successful candidates.

The following explanations are given as information to those who intend to take the examination.

Evaluation will be conducted with the utmost care through a number of processes, which are outlined below:

1) Individual Evaluation of Common Questions:

   All examination papers are marked by each examiner (Member of the Board of Examiners) based on the criteria to be hereinafter shown and are then submitted to the Board through the Examiner-in-Chief for accreditation.

2) Evaluation of Papers Regarding the Industrial/Scientific Divisions:

   Examiners in charge of respective industrial divisions will mark the papers based on the established criteria and point system. The results are reported to the Examiner-in-Chief. The marks thus given are reported independently without considering the marks given in the Common Questions test.

3) Final Evaluation by the Joint Committee:

   The Examiner-in-Chief will convene the joint committee for the final evaluation and grade accreditation. The committee consists of 19 Japanese members and several foreigners.

The National Board of Examiners for
Foreign Language Proficiency Test
in Science & Technology

CRITERIA ON MARKS AND GRADE ACCREDITATION IN EVALUATION

Overall Criteria in Evaluation: Grade Accreditation Criteria (Common to All Languages)

Grade	Degree of Accomplishment and Applicable Standard
1st Grade	Translators in this category should have an excellent knowledge of the Japanese language in respect of its structure and versatility regarding expression. They should also have competence to express what the source text means in a corresponding foreign language (target language) with accuracy, correctness, and naturalness, thus making the translation easy to understand for foreign readers. To be more specific, each translator should be equal to the task of analyzing a given Japanese source text for immediate conceptional conversion into a style suitable to the particular output language The correct translation should be completed in a reasonably short time, as specified in each section of the Test and must be easy to understand by native speaking people. A good style in addition to terseness, consistency, accuracy and correctness are the essential qualities for translation-type writing.
2nd Grade	Translators of this category should have a good knowledge of the Japanese language and must be able to transfer what is meant in the source language text (Japanese language text) to a target language and must be easily read by native speakers without misunderstandings due to differences of expressional styles found in the two languages. Competence to make documentation in a foreign language consistently and naturally, without being influenced by any style peculiar to the source language (Japanese), is also an important quality.
3rd Grade	Translators who belong to this Grade have been so rated because they do not satisfy the requirements for the 2nd Grade but show promise provided that they improve themselves, especially with respect to competency in translating the source language (Japanese) text in spite of any style difference between the two languages.

3rd Grade (cont.)	The standard for this Grade is a competence to correctly translate original Japanese papers into a target language notwithstanding some extent of unnaturalness in the ways of expression or the selection of terms.
4th Grade Associate LTT	Although endowed with an aptitude for correct translation, those who come under this category are required to make further improvements regarding their competence before their translations into a target language are acceptable for practical use. However, they have been so rated in recognition of their potential for obtaining the qualification of a Licensed Technical Translator.

NOTE: The final judgment is made based on the integration of all the examiners' evaluations in accordance with the detailed marking standards (unpublished guidelines for the Examiners only).

These standards are applicable to all languages and industrial divisions in the Test without modifications.

DESIGNING EFFECTIVE LANGUAGE TRAINING FOR PROFESSIONALS

Galal Walker

Two years ago, the Chinese Language Program at Ohio State began to accommodate a new group of students in its elementary courses. We refer to this group as "non-traditional students," a category that comprises several kinds of students not ordinarily found in our language classrooms. At Ohio State, the non-traditional student might be someone from a college that does not have a language requirement -- such as the College of Engineering -- or, perhaps, a student who is pursuing a major in a pre-professional academic program that does not allow sufficient time to follow a classroom course in a difficult language. Non-traditional students also come from off-campus. Enrolled through our College of Continuing Education, these are the students who for the most part populate the evening courses and courses offered by Ohio State's Weekend University. Although most of our non-traditional students are people who are learning Chinese as part of their preparation for future careers, the College of Continuing Education has sent us a significant number of men and women who are already established in their careers and are embarking on the study of Chinese for specific professional purposes.

These professional people in our beginning Chinese courses are a heterogeneous group: We have businessmen who are currently involved in commerce with China or Taiwan or who have specific plans for doing so in the near future; there are engineers who plan to work on projects in China; some are teachers who will go to China as participants in educational exchanges; one is a manager at a local computer-oriented company who is learning Chinese in order to demonstrate an interest in his colleagues from Taiwan; some are government workers, and in addition, we have a lawyer, a bank manager, and a roughneck who expects to be working on an oil rig in the South China Sea next year.

Some of these students have limited goals for their study of Chinese. They want no more than to be able to accomplish some basic tasks when they are in China and to be able to demonstrate a polite interest in China and the Chinese language for their Chinese counterparts. Others are in it for the long haul and appear to be committed to a period of study sufficient to become functional in Chinese society. Their backgrounds in foreign languages are extremely varied: A few speak more than one foreign language and demonstrate an admirable sophistication in the way they set about learning Chinese. Unfortunately, however, most have never learned to speak a second language and do not have the vaguest notion of language training. Were it not for the specific direction and redundant presentations in the specially designed materials used in the courses, most of these inexperienced learners would not be able to begin to acquire the most basic language skills.

The diverse group of professional people who have set upon a course of Chinese language study are, however, united by one dominating characteristic -- they all have strict constraints on their time. Each of their schedules is unique and their time is valuable. Most of them can quote a price for their time by the hour or day. This places two requirements for the program that is designed to accommodate these people: it must be able to deliver instruction according to the students' schedules, and it must waste no instructional opportunity.

Instruction in Chinese is provided to these non-traditional students under the aegis of Ohio State's Individualized Instruction Program. Individualized Instruction per se has been amply described elsewhere.[1] The purpose here is to relate how this type of instruction works for Chinese and to discuss the resources that are necessary to deliver the instruction while maintaining the high quality that makes the program worthwhile for both student and teacher.

People who are primarily occupied with establishing and advancing professional careers are not able to arrange their lives according to the semesters or quarters of the university, nor can they long maintain the schedule of a regular classroom course. Business people have cycles of high and low activity and intensive traveling. The roughneck in our program at Columbus, for example, spends three weeks off the shore of Louisiana and three weeks back home in Columbus. The provider of instruction to such individuals must be able to adjust the pace of the training to each person's available time. Thus, the courses must be self-paced. This poses special problems for the study materials, the instruction, and the administration of the courses. The simplest problem to solve is that of administration, although some administrative offices may not have the flexibility to accommodate an individualized instruction arrangement. Several educators who inquired about establishing individualized instruction courses for non-traditional students terminated their inquiries when it was made clear that their administration would have to enroll students and record their achievements on the basis of variable credit hours. Their deans and/or their computers were not sufficiently advanced to deal with this untidy arrangement. It should be clearly understood, therefore, that if administrative procedures do not permit variable credit courses, there is no practical way to provide an extended course of language training for people concurrently involved with professional careers.

At Ohio State, the administrative procedures for such self-paced courses were established years ago and now function as a matter of course. A student in a ten-week Chinese course can register for from one to five credit hours. During the sixth week of the course, the student assesses her progress to date and makes a contract with her college to complete a specific amount of the course. If the amount to be completed differs from the initial enrollment, she adds or drops hours and her obligation to her school, financial or otherwise, is likewise adjusted. Thereafter, as long as the student responsibly fulfills this contractual requirement, there is no problem. If the student fails in this regard, the problem is his/hers.

The administration must also provide a continuously available space for the program. With each student in the program working at an individual pace and completing different amounts of the coursework, the traditional classroom instruction is impossible and there cannot be a regular schedule of classes. Our program has been assigned a modest space that we have designated the Chinese Individualized Instruction Center. This center contains files of each student's records, two language laboratory positions, our audio tapes, a high-speed tape duplicator, and a couple of tables. It is in the center that all face-to-face contact between the instructor and individual students occurs. The student is provided with a set of self-instructional materials that includes an agenda. The agenda, a set of instructions on how to proceed through the materials, informs the student about what to study, in what order, and at what point she is to see her instructor. Times for meetings with the instructor are arranged by the student. The center is staffed for approximately 25 hours a week, which includes evening hours as well as three hours on Saturday. The student may come to the center any time it is open and practice at the language laboratory facilities, copy tapes, or work with an instructor when he is free. In order to avoid waiting for the instructor on duty to finish working with another student, if the student wants to meet with a particular instructor, such time can be reserved either by signing up in an appointment book or by telephoning the center ahead of time.

At the time the appointment is made, or at the beginning of the appointment with the instructor, the student informs him of the purpose of the meeting. We have two basic types of meetings: FACT sessions and ACT sessions. FACT sessions are conducted in both Chinese and English and focus on knowledge about the language -- questions on grammar, articulation, and the social context of the language presented in the study materials. FACT sessions are proportionally few. More often the student will request an ACT session, which is conducted in Chinese and is conceived of as a performance of what the student has been studying. In ACT sessions, the student demonstrates the ability to use the material just studied. This ability is demonstrated in a set format. First the student performs a short dialog with the instructor, who uses the dialog to evaluate pronunciation and intonation. When the dialog is performed satisfactorily, the instructor engages the student in exercises that require a command of the newly studied material and of specific review material. A satisfactory performance of the dialog and exercises earns the student the right to continue on to the next stage of the course. If performance is below standard, weaknesses will be explained and the student will be told to re-study the lesson and, if necessary or desirable, to come in for an additional practice session before again trying to demonstrate command of the material.

Meetings between student and instructor generally last from ten to twenty minutes. Since the business of each meeting is clearly delimited by the agenda and the content of the lesson being exercised, a well-prepared student who has no difficulty with the material can be checked out in five to fifteen minutes of intensive exercises. As soon as an ability to use the material in the lesson is demonstrated, the student has the option of terminating the meeting or joining the instructor in a language activity related to recent and past lessons. Of course, the further along in the course a student is, the longer the meetings tend to be, since the student can do more with the language. By engaging the instructor in fifteen to twenty minute sessions of intense language activity, the student in an individualized instruction course will have more language exchanges with her instructor per unit than will a student in a regular classroom course.

Grades are not a motivating factor in the language training of professional people. When a person begins to discuss the possibility of learning Chinese in our program, she is sure to explain that she does not have the slightest concern about her transcript and what goes on it. As a means of quality control, a traditional grading system is of minimal use in a course involving people who are not looking for a job. At Ohio State, individualized instruction courses have adopted a "mastery-based" system of evaluation. The minimum passing grade is 80 percent, or B-. In the Chinese program, a B- represents a level of performance that is acceptable to an indifferent native speaker of Chinese. Unless the student is demonstrating a capacity in the Chinese writing system, all performance evaluation is by means of oral testing. Students who persist in a particular grammatical or pronunciation error may be given a supplementary written quiz as a means to focus on pertinent facts; however, whether a student is to advance in the course or not is determined only by oral performance. When the student's performance convinces the instructor that her Chinese would work with a Chinese person who is not inclined to make extraordinary efforts to communicate with a foreigner, the student has won the right to continue on to the next level.

Once college-level administrative procedures for accommodating variable credit courses are in place and space has been made available to house the program, the next level of problems regarding this kind of instruction may occur within the department. The department allocates instructional resources, and it is in the department that the work will actually be done. Language courses are always underfunded -- and the language environment in which our courses are conducted is grossly impoverished. A psychology laboratory is able to employ more equipment to teach a rat to turn left rather than right than we have available to teach a group of promising young students to communicate in Chinese.

I know of no language course that cannot put additional resources to good use. When a department decides to provide for this new constituency of language learners, it has the painful choice of either directing needed additional resources away from its classroom courses or thinning out the already scarce resources for its traditional courses.

What makes the decision even more difficult is the fact that the professional people who receive this instruction do not have the impact on a department or program that the traditional students have. Non-traditional students are usually involved with their courses when everyone else is away from the department; they do not attend literature or culture courses; they do not socialize with the rest of the department; and they do not join the list of majors. Thus, the need to provide this service is not a daily concern of the various members of a department. The decision to provide specialized instruction for these people is based either on the hope that there are sufficient numbers of these students to make the effort worthwhile, or on the conviction that this group of language learners includes individuals who are especially important people to reach.

Although in our regular language courses at Ohio State there is a clear increase in the number of students engaged in technical and professional fields of study, there are no verifiable data that identify a substantial number of professional people with the need and desire to learn Chinese. Even though it is clear that the professional people among our non-traditional students are the ones who will immediately put their language training to use, the conviction that these are important people to provide service to is moderated by the knowledge that providing elementary language training to people late in life, as we do with these professional people, is, in the broader mission of a language department, a stop-gap measure. Ideally, training in a language as crucial as Chinese should be sufficiently broadcast in the academic community to include a good number of future engineers, chemists, lawyers and the like. That must be the direction of growth for any responsible language program. And ideally, professionals with career needs for Chinese or any other language need to gain language proficiency early so they do not have to combine their career activities with as low a level of proficiency as will result from a belated start in the language.

The present situation, however, falls short of the ideal state. There are such people who need to begin to learn to use Chinese. Having made the decision to provide language instruction to these people, the Department of East Asian Languages and Literatures at Ohio State currently allocates three graduate teaching assistants, working under the supernumerarily assigned supervision of a faculty member, to provide instruction to 35 students. If the graduate assistants are trained, this is an adequate, even generous teaching staff. However, there is an almost constant need to train newly assigned graduate assistants and, since an instructor in an individualized instruction course must be able to teach any part of the material in four courses, the training is extensive and time-consuming.

At this point in the development of Chinese individualized instruction, greater teaching resources are necessary. The curriculum is still in the process of development. In order to sustain and advance that development, an interested and capable faculty member should be given the task of developing the curriculum and promoting the program as a regular assignment. At this stage, maintenance-level staffing is premature.

At Ohio State, without any sort of promotion or recruitment activity and with native and non-native-speaking graduate teaching assistants who have received sufficient training to keep the course going, our program continually attracts 25 to 45 students. With promotion and more permanent staffing, the size of the program could easily be doubled. In a more favorable location, say a downtown campus in a major city, the potential for this type of program appears to be great. Whether or not this potential can be realized depends largely on the ability to meet the curriculum challenges presented by this instructional mode. The development of appropriate teaching methodologies and the development of instructional materials are equally crucial to this matter.

Despite the one-to-one contact between teacher and student, the role of an instructor in individualized instruction is neither that of native informant nor that of tutor. The individualized instruction teacher needs to be able to give short, fast, intense, and effective lessons over a wide range of materials. Whether or not the teacher is a native speaker, he needs to be very familiar with the instructional materials of the courses and to have enough pedagogical sophistication to understand and deal creatively with the uniqueness of the situation. Of course, experience and technique gained from teaching in the classroom have applications in the individualized instruction session; however, there are significant differences. The group dynamics of a classroom can sustain activities that are not possible when there are just two people talking. Rapid-fire intensive question drills on a set grammatical pattern can be an effective and interesting class activity, but in a one-to-one setting they are more likely to be merely irritating. The vast majority of methodologies and techniques available to the field are designed for classrooms. The individualized instruction teacher must be able to create an appropriate methodology by sorting through these and adapting them to his situation. At present, we are at the beginning stages of this process.

The social role of the instructor in individualized instruction is different from that in classroom instruction. The teacher-leader of the classroom is replaced by the experienced adviser in private sessions. This different role may be disputed by some skilled classroom teachers who view one-to-one instruction as a less prestigious activity than directing the activities of 10 to 15 students for 50 minutes. In Chinese, this presents a special problem. Since Chinese culture tends to give greater status and attention to teachers, the role of the teacher is more defined. Problems that we have had with ineffective teachers have often derived from a native teacher's inability to take the individualized instruction as seriously as he would the instruction in a classroom setting.

The critical factor in providing quality instruction is the instructor who is trained to exercise the single student effectively, to use teaching materials efficiently, and to adhere strictly to evaluation standards. In this instructional mode, where the instructor is the sole arbiter of standards, any chronic negligence on his part can mislead the student into thinking she has skills which she in fact does not possess -- a situation that results in wasted effort for all involved.

Record-keeping is a very important feature of this kind of instruction. The results of each session must be noted in the student's record in sufficient detail to enable the next instructor to determine quickly where the student is in the course and what special problems she may be having. Records of this sort are not commonplace in classroom instruction. Therefore, each new instructor must be carefully familiarized with the record-keeping system and must be convinced of the critical role the student records play in the execution of the course.

In this program, the student's contact time with the language is predominantly spent in self-study. This places a greater burden on the instructional materials than is the case with classroom instruction. Whether the materials are text, audio tape, video tape or computer-assisted instruction programs, they must be readily accessible to even the unsophisticated learner and must be complete enough to be used without constant consultation with the teacher. At Ohio State we are in the process of compiling materials specifically for this mode of instruction. Our goal is to include in our tapes and texts everything that would ordinarily be presented to the learner by the classroom teacher. The experience has reinforced our suspicion that Chinese teachers more than earn their pay. Compared to the commercially available textbooks used in our classroom courses, the individualized instruction materials are four times more voluminous. We target our materials toward what we have determined to be the average consumer of these courses, not toward the optimum language learner. Our average student is an older, busier person who has never learned to communicate in a foreign language. For this person we have created materials that are more redundant in information and include many drills that are more varied and numerous than the drills in any of the standard Chinese textbook and tape program combinations. For the more capable or more sophisticated learner, these materials may be excessive. However, we have found that it is more effective to have more rather than less. The abler learner can skip unneeded material, but the less capable learner cannot herself provide necessary information or drill when it is lacking.

Almost any instructional technology can be applied to individualized instruction. However, we must be sure to employ the technology that is appropriate for our purposes. The most effective tool for learning to speak Chinese, in our experience, is the audio tape. It is the most convenient for our students, who commonly study by working with audio tapes in their automobiles as they commute to and from work. Therefore, our materials focus on a program of audio tapes backed up by a workbook text. More advanced technologies such as video, computer programs, and interactive video surely will have a future role in this sort of course, but the priority for now is tape and text.

Aside from commercial language schools that provide personal tutors, individualized instruction is the only institutionalized form of language training available to the professional person. If conducted with care and pedagogical integrity, it can be effective. It is, however, not a quick fix, nor is it cheap or easy. Once it is in place, however, it can be a unique and valid way to meet the need for language training.

FOOTNOTE

1. Twarog and Walters, "Mastery-based, Self-Paced Instruction in Foreign Languages at Ohio State," Modern Language Journal 6.5.1., pp. 1-23.

PROBLEMS IN TEACHING TECHNICAL JAPANESE

R. Byron Bird and Edward E. Daub

In discussing the subject of teaching technical Japanese, one is immediately faced with two approaches to the subject: a beginning Japanese course followed by a course in vocabulary building and comprehension and translation techniques for technical Japanese; and a restricted course aimed exclusively at the "decoding" of technical Japanese. Instructional materials have been prepared for both approaches. At the University of Wisconsin we believe firmly in the first approach and have prepared a technical Japanese reader appropriate for this teaching method.[1] On the other hand, at the University of Sheffield in England[2] and in East Germany,[3] teaching aids have been developed to support the second approach.

By a "beginning Japanese course," we mean a two-semester course that includes an introduction to Japanese grammar, practice in elementary conversation, thorough mastery of <u>hiragana</u> and <u>katakana</u>, an introduction to the writing and reading of <u>kanji</u>, and the development of facility in using a <u>kan'ei-jiten.</u> At the University of Wisconsin, the first-year course consists of two semesters, each for six credits; this includes conversation practice and use of tapes for developing aural comprehension. Currently the textbook by Mizutani and Mizutani[4] is being used.

Our reasons for strongly preferring the first approach mentioned above are as follows:

1) Learning the spoken language helps the student to develop a feeling for the structure of the language and to acquire the intuition which is so vitally needed in order to read for comprehension (as opposed to "decoding").

2) Technical Japanese translators and practicing engineers and scientists will ultimately want to meet and talk with Japanese colleagues, and the beginning Japanese course can provide the basis for developing the ability to do so.

3) Japanese teaching programs are continuing to develop new teaching methods and teaching materials that make it possible for beginners to acquire an excellent foundation in the language prior to embarking on the more specialized task of technical translation.

4) Technical Japanese sentences are often very long and complex, with many subordinate clauses and formal constructions; these involved sentences are very difficult to comprehend unless one has had considerable experience in handling short simple sentences.

It might appear that, if one wishes to do scientific Japanese translation only, a "crash course" aimed directly at that narrow goal might be more efficient and less time consuming. However, we feel that the solid background provided by a beginning Japanese course ultimately results in a far better ability to read and comprehend technical material -- as opposed to laborious "decoding" -- and is therefore well worth the additional time required.

Problem 1: Students Must Learn Basic Japanese First

Not all American universities provide a beginning Japanese course. However, this is not an insurmountable problem. Middlebury College in Vermont offers a summer intensive course in beginning Japanese. In addition, the Mizutani and Mizutani text[4] might possibly be used for self study, inasmuch as the six cassettes that accompany the book provide an adequate introduction to the phonetics of the language. The textbook of Jorden,[5] well known for its clarity, accuracy, and sound pedagogy, is also very well suited for obtaining a thorough grounding in the essentials of the grammar by self study; it would have to be supplemented by additional study materials on _kana_ and _kanji_.

A beginning Japanese course develops skills in many areas that are important in technical Japanese. Let us summarize some of these and show why they are important later to those who wish to learn how to comprehend scientific and engineering texts.

To begin with, one must know how to pronounce the basic sounds in Japanese and how they are written both in _romanji_ and in _kana_; this is necessary for the use of various kinds of dictionaries, some arranging the entries in the ABC system with words written in some kind of Romanization, and others using the AIUEO system with words written in the Japanese syllabary. Romanized texts are frequently used in beginning courses, and they are in some ways particularly appropriate inasmuch as the student can learn more quickly how to identify various parts of speech and to recognize the individual words as distinct entities. The _hiragana_ and _katakana_ must, of course, be mastered at an early stage in view of their fundamental role in the normal written language. Scientific texts contain many foreign loan words (_gairaigo_) such as _metanooru_ (methanol) and purasuchikku (plastic), and these are invariably written in katakana.

A beginning Japanese course also provides an introduction to kanji, the basic ideographs used for writing most nouns, adjective stems, and verb stems. Knowing how to count strokes and identify radicals is essential for efficient use of a character dictionary (kan'ei-jiten), and such a dictionary must be frequently consulted in doing technical Japanese translation. It is also very important that the aspiring technical translator have a good understanding of the "on" and "kun" readings, inasmuch as he will often encounter character combinations (jukugo) that are not to be found in the kan'ei-jiten, and it will be necessary for him to make a shrewd guess as to the pronunciation of the component kanji in order to look the word up in a Romanized dictionary. The beginning Japanese course should provide him with the necessary insight.

The introductory course in Japanese generally provides an explanation of the most important inflected forms for verbs and adjectives. Because of the great regularity in the language, the learning of these forms usually does not present a problem; on the other hand, acquiring a feeling for the precise translation of the various inflected forms requires rather careful indoctrination, and this material is vital in scientific translation.

Of particular importance in the beginning stages of learning Japanese are the "particles," which signal the function of various words in the sentence and their interrelation. These little guideposts do not have precise equivalents in English and therefore require extremely careful teaching and learning. Equally important are the "noun-following expressions" and the "verb-following expressions" (such as bakari de naku, koto ni natte iru, and wake ni wa ikanai) which pervade the language and create enormous problems for the novice. The beginning Japanese class can provide a solid introduction to these aspects of the language. A particularly nice summary of these features of Japanese has been provided by McClain[6] in two small volumes, which can be particularly helpful to the fledgling scientific translator.

Of crucial importance in understanding Japanese is the development of a thorough understanding of the structure of complex sentences, in particular those with several subordinate clauses. A beginning Japanese course generally helps to develop the feeling for how complex sentences are put together, and the technical translator must develop the techniques for unraveling extremely long and complicated sentences. Scientists seem to revel in this kind of writing, either because they are attempting to express intricately connected thoughts or because they are poor writers and don't seek the most economical and concise way to express themselves.

These, then, are the aspects of the language that we feel should have been studied in an introductory Japanese language course before one proceeds to the study of technical Japanese translation: the basic sounds of the language and their transcription, the structure and readings of *kanji*, the endings used for verbs and adjectives, the use of particles and verb-following expressions, and the structure of complex sentences. Above all, the beginner should have acquired a feeling for the language, its peculiar features, and its charm.

At the University of Wisconsin, the course in technical Japanese has been taught to students with a variety of backgrounds. Those who have had the one-year beginning Japanese course have been able to move into the course based on our book with a reasonably easy transition. These were students in technical areas who had a very solid grounding in scientific subjects and hence were already somewhat familiar with the topics discussed in our textbook. On the other hand, one student who had had four years of Japanese reported that the use of particles in complex sentences still provided a major obstacle.

By way of an experiment, we attempted to teach the technical Japanese course to graduate students after giving them a short course of 30 one-hour lessons. This short course, providing some training in basic grammar and conversation, was based on the NHK radio course[7] which was broadcast on the Wisconsin Educational Radio Network. The radio program was supplemented with *kana* exercises and classroom drills on complex sentences with modifying clauses.

One graduate student in chemistry, upon completing this short course, then took the technical Japanese course, but she found it difficult to comprehend complex sentences since she had not had enough experience in coping with sentences containing modifying clauses and phrases. On the other hand, she was often able to make reasonable translations simply because of her understanding of the scientific content. This is an important point -- that persons trained in science have some advantage in doing scientific translations simply because their knowledge of the subject material enables them to overcome some of the linguistic problems. This brings us to the second major problem in teaching technical Japanese: the myth that technical Japanese is hopelessly difficult.

Problem 2: Is Technical Japanese Hopelessly Difficult?

Once one has had an introductory year course in French, German, or Russian, it is a relatively simple matter to develop facility for reading scientific texts in these languages. The principal hurdle is the acquisition of the necessary vocabulary of technical terms and scientific jargon. Many textbooks have been prepared for helping scientists and engineers master the art of technical translation of French, German, and Russian, but for the most part these are merely collections of graded texts with an accompanying vocabulary list. Does technical Japanese pose some particular problem not encountered in the European languages?

As a matter of fact, technical Japanese is considerably simpler than one might expect, particularly compared with conversational Japanese. To begin with, there are quite a few topics covered in the beginning Japanese course that do not arise in texts in science and engineering:

1) The "-masu" forms of the verbs are almost never encountered in technical writing, except in some operating instructions for apparatus and equipment.
2) Keigo (polite language) is never used, so that humble and exalted verbs never arise, nor do honorific prefixes and suffixes for nouns; furthermore, the various nouns designating family relationships and the many expressions involving the verbs of giving and receiving never arise.

3) One never encounters the various abbreviated forms used in conversation, such as "-chau" (for "-te shimau"), "ja" (for "de wa"), and "tte" (for "to itta").

4) The suffixes used for "counters" are seldom seen in technical writing.

These various topics are of great importance in spoken Japanese and are, of course, also necessary in reading novels and short stories.

Those who wish to emphasize technical reading do not need to be particularly skilled in oral expression or in aural comprehension. One can enjoy the luxury of working at one's own comfortable pace in reading, comprehending, or translating the scientific or engineering text. Thus a somewhat limited set of skills is required.

In reading literary Japanese, one encounters a variety of problems because the men and women of letters enjoy the delights of language for their own sake. They like to savor the flavor of words, use abstruse language for literary effects, and amuse themselves and tease the reader with vague and mysterious phrases. Reading some poetry and prose often requires a superb knowledge of the details of the language and, in addition, an understanding of psychology, history, social customs, and religion. On the other hand, reading scientific and engineering materials is relatively straightforward.

Technical writers are supposed to avoid ambiguity and present facts, figures, theories, hypotheses, and conclusions in as straightforward a manner as possible. Generally speaking, there is a one-to-one correspondence between the Japanese and English words that designate the concepts, objects, instruments, scientific laws, machines, and symbols that are discussed in scientific texts. If one is doing a translation in a subject that is familiar, the task is particularly easy. In fact, knowing the scientific material can guide one through the difficult structure of a complex sentence.

Technical Japanese is also easier than literary Japanese because a much smaller list of _kanji_ is required to gain access to the published literature. Whereas the humanists and social scientists have to deal with the full gamut of human emotions, motives, ambitions, and hopes, the scientist concerns himself with a rather restricted world inhabited by chemical elements, elementary forces, laboratory apparatus and equations. Moreover, the _kanji_ needed in reading technical Japanese are far fewer than are needed in reading Japanese literature, Buddhist texts, or newspapers.

We see that in many ways technical Japanese is far easier than conversational or literary Japanese. Why then do we hear that technical Japanese translators are in such short supply? Why have science and engineering students not been studying Japanese so that they will have access to the technical literature that reflects the stunning advances being made in Japanese industry, and in university and government research laboratories? It is partially because traditional Japanese language programs have not developed materials designed to teach technical translation. This led us to the preparation of our textbook, _Comprehending Technical Japanese_.

Problem 3: Need for Specialized Teaching Materials

Anyone confronted with the task of learning Japanese inevitably asks: Which _kanji_ should I learn first so that I will be able to recognize the most important words in the language? The answer to this question depends on the type of reading material that the person expects to encounter. We attempted to answer this question in 1967 for the field of physics[8] by compiling a frequency count for the _kanji_ (43,756 of them!) that appeared in a high-level Japanese high school physics textbook. The top 12 characters on the list accounted for 20 percent of the _kanji_ in the book, the top 60 accounted for 50 percent, and the top 300 covered 90 percent.

Later Professor Nobuo Inoue of the Science University of Tokyo, together with some of his students, made similar frequency studies for chemistry and biology, also using high-level high school texts. From these three lists, we then compiled a master list of 524 _kanji_, which included all characters that appeared in the top 300 of any one list, as well as those that were in the top 400 of two or more lists. These studies suggested that mastery of about 50 characters would provide a scientist or engineer with a very good basis for reading technical material. It seemed to us that a sensible project would be to prepare a technical Japanese reader featuring 500 characters. (We actually used 490 of the top 500 characters plus ten others to make up the list of 500 for our reader.)

One might question our use of high school texts for doing the frequency counts, but these books, written by top-notch Japanese scientists, cover a very wide spectrum of topics. It was our belief that frequency lists based on these books would provide a reasonable starting point for foreigners interested in mastering technical Japanese.

It is interesting to compare our list of 500 characters with the list of 500 compiled by Crowley;[9] he combined two studies published by the National Language Research Institute of Tokyo[10], based on "cultural reviews." (Crowley points out that "general magazines" might be a more appropriate description of the text material analyzed.) We found that only one-half of our most frequent _kanji_ were included among the most frequent 500 in Crowley's tabulation.

More significantly, perhaps, we discovered that 60 characters in our science-based list not only did not appear in his top 500; they did not even merit a ranking in his tabulation of the _toyo kanji_. Apparently they did not appear with any frequency worth recording in the National Language Research Institute data base of general magazines. Among those 60 were the characters for liquid, solidification, reactor, nucleus, acid, oxidation, bacteria, radius, flame, expansion, copper, crystal, membrane, and equilibrium -- all concepts fundamental to the sciences.

Another comparison can be made with Kikuoka's tabulation of the 1,000 most important character compounds (_jukugo_) appearing in newspapers.[11] In the top 100 _jukugo_ of Kikuoka's list, 25 compounds appear in which neither character is on our list of 500. These include words such as _iin_ (committee member), _seifu_ (government), _senkyo_ (election), _kokumin_ (people, nationals), and _kokkai_ (parliament) -- words which would seldom be encountered in scientific material.

These comparisons of frequency studies emphasize the need for special teaching materials for students of science and engineering. Most of the traditional Japanese courses and textbooks emphasize those _kanji_ that are crucial to those who intend to study Japanese literature, social sciences, and religion. Even if a scientist or engineer takes three years of Japanese in the standard program, he will not be adequately prepared for the first technical translation task he encounters.

When the two of us joined forces in 1971 to prepare our book, we were delighted that Professor Nobuo Inoue of the Science University of Tokyo was willing to join us. He was well prepared to take part in this project, since he had spent a total of nine years in the United States and England, including a professorship at Purdue University. We admired his amazing facility with English and his talent for writing excellent Japanese textbooks in physics and mechanical engineering.

Our book has a total of 25 chapters (the first 17 devoted to physics, the next five to chemistry, and the last three to biology). Each chapter includes the following material:

1) A list of twenty kanji with the most important "on" and "kun" readings needed for technical reading.

2) The vocabulary list for the main reading selection.

3) The reading selection (first in kanji and kana, then in romanji, and finally in English translation); this reading selection features the twenty kanji given at the beginning of the lesson.

4) Explanatory notes to assist the reader with difficult passages and to provide translation aids; also supplementary information on how to "read" equations and formulas in Japanese.

5) Construction examples to emphasize and illustrate certain grammatical constructions that arise frequently in technical Japanese; these are all constructions that appear in the main reading selection.

6) Supplementary readings, with vocabulary lists; these readings, taken from a wide variety of sources, were selected to emphasize the twenty kanji featured in the lesson.

7) Final translation test, which contains no vocabulary list or explanatory notes; the reader is encountering a "real-life" translation experience here.

Each lesson, then, features twenty kanji and their most important readings, and the reader is expected to master these twenty and be able to recognize them in all subsequent chapters. Occasionally characters are encountered that have not been featured in the current or foregoing chapters; these are provided with furigana in order to indicate the pronunciation. In this way the student sees the text exactly as it appeared in the original source, aside from the appending of furigana on those charactes that have not yet been mastered.

We feel that the "Construction Examples" are an important feature in each lesson. These are frequently occurring expressions that often create difficulties for beginners, such as "...ni hoka naranai" (is nothing but...); "...ni motozuku" (based on); "...ni zoku suru" (to belong to...). Professor Inoue wrote example sentences to illustrate these constructions, using the kanji and vocabulary of the main reading selection. He wrote these in very clear and direct style so that the student could see further examples of the construction without finding it buried amidst other complexities.

We did not make a formal study of the frequency of these grammatical constructions. By the end of the book, however, few new constructions presented themselves, and therefore we have reason to believe that our list is fairly complete. Professor Akira Miura, who heads the Japanese language program at the University of Wisconsin, reviewed our list of construction examples and found that about 20 percent of these are covered in the first-year course, about 36 percent arise in the second-year course, about 22 percent in the third-year course, and about 22 percent beyond. Thus, in the normal curriculum, some 40 percent of the constructions that occur frequently in technical Japanese are not covered at the end of the first two years. This is a further example of the need for special instructional materials for scientists and engineers.

Scientists and engineers and professional technical translators want to learn to read a foreign language and understand what they are reading. They also must be capable of making accurate and responsible translations. Acquiring this facility requires much practice. In each lesson we give the student the opportunity to gain experience with varying degrees of assistance: first in the main reading, where much help is given; then in the supplementary readings, where vocabulary lists are supplied to free the reader from having to spend a lot of time looking up words in the dictionary; and at the end, the final translation test when the student is entirely on his own. These final reading selections were written by Professor Inoue, some as original essays, others as edited texts from a variety of sources.

There are still many projects that need to be undertaken in connection with technical Japanese translation; for example:

1) preparation of a companion volume to our book aimed at helping Japanese language teachers use it for tutoring science and engineering students in independent study or in the classroom;

2) preparation of frequency lists for various subfields of the sciences, such as quantum mechanics, rheology solid state physics, macro-molecular chemistry, laser technology, and operations research;

3) preparation of supplementary lessons for use with our book to help people learn the most frequent _kanji_ and _jukugo_ in these various special subfields. (We are currently laying plans to do this.)

Clearly there are many lines of activity that could and should be pursued in this regard. However, nothing much will be done without the financial support of granting agencies or industry. Probably such support will not be forthcoming until there is more widespread recognition of the need for special educational materials. This is the final problem that we see in teaching technical Japanese: the institutional barriers to providing the resources for language instruction.

Problem 4: Institutional Barriers to Teaching Technical Japanese

Up until about 1970, practically all universities required that Ph.D. candidates pass reading examinations, generally administered by the Graduate School, in two foreign languages. It was felt that anyone entering on a research career should have access to the foreign literature in his own field. In the midst of the upheaval of the late 1960s and early 1970s, these language examinations were abolished, and in some institutions it was left to the individual departments to decide on whether or not foreign language competency was needed. Now, 15 years later, departments requiring foreign language competency are rather rare. As a result, there is no encouragement for students in the sciences and engineering to study languages for use in research. Furthermore, very few professors are themselves able to read research papers in foreign languages, and therefore students do not feel any particular obligation to concern themselves with published research if it is not in English.

In addition, the language requirements for the B.A. were lowered about 1970. Pre-university training in languages is abysmal; in the state of Wisconsin, for example, less than 2 percent of graduating seniors have had four years of one foreign language. Thus, throughout the entire educational system, foreign language training has been de-emphasized. As a consequence, there has been little or no demand for the teaching of technical Japanese.

Until very recently, the accrediting agency for engineering curricula (the Accreditation Board for Engineering and Technology) would not allow credits in introductory foreign languages to count as "socio-humanistic studies" in calculating the credits required for graduation with a B.S. in engineering. The argument was that introductory foreign language courses do not have sufficient cultural content. As a consequence, engineering students have not been able to study foreign languages without adding ten to twelve extra credits to their already demanding course load. Now that the accrediting agency has changed this rule, perhaps we will see some increased interest in foreign language study by engineers.

When universities had required reading examinations for a Ph.D., it was not uncommon for courses in scientific German, French, and Russian to be taught by the language departments involved. After the scuttling of the language examinations, the demand for these courses dropped precipitously and then the courses ceased to exist. Now most language departments are reluctant to institute such courses again; in some institutions, scientific German, French, and Russian are offered as a "University Extension" course outside of the main university.

Most foreign language teachers do not wish to teach technical translation courses, inasmuch as they feel uncomfortable with the subject material and the teaching of such courses would do nothing to further their own scholarly progress. Another possibility is to have the technical reading courses taught by professors from science and engineering departments; but such persons would not have the necessary credentials in linguistics and would therefore be unacceptable to the language departments. Therefore the field of technical translation has, in fact, become a stepchild in academia. Little or no encouragement is being given to the teaching of technical Japanese.

One very real problem in getting science and engineering students interested in foreign language study is the fact that most first-year language courses are very time-consuming. At many universities, the first-year Japanese course consists of five or six credits per semester and since science and engineering curricula are already very heavy -- often with a heavy dose of required laboratory courses -- it is virtually impossible for students to take beginning Japanese. If the Japanese language staff were large enough and if the enrollments in Japanese were sufficiently large, it would be very fine for the science and engineering students if a sequence of three-credit courses could be made available. Until that happens, we do not expect to see any significant increase in interest in technical Japanese among students in engineering and the sciences.

It is our impression that in industry there is very little interest in job applicants who have a foreign-language capability. No extra reward is given for those who can do technical reading in foreign languages. Even companies with joint ventures in Japan do not seem to be particularly interested in trying to hire technical people who can communicate in Japanese. It is generally just not regarded as a problem. The "let-them-speak-English" attitude prevails. For years, the feedback among students headed for industrial jobs has been that there is no point in wasting time on foreign languages. Here again we find actual discouragement for anyone who might have an inclination to study technical Japanese. Most industrial research and management people do not seem to be concerned; apparently they can get technical articles translated when necessary, and they claim that abstracting services are adequate for their purposes.

We realize that this conference represents a change in climate for the teaching of technical Japanese and that the MIT Workshop on Japanese Scientific and Technical Information in January 1983, as well as the Congressional Hearings on Availability of Japanese Scientific and Technical Information in the United States in March 1984, identified many of the problems we have mentioned, especially the lack of attention to the needs of science and engineering majors in traditional Japanese language programs. The lead article to the Workshop Proceedings describes the situation very well:

> "...science and engineering majors who might wish to add Japanese language to their curriculum face the fact that most of the language programs found in the U. S. universities are targeted not at science or engineering students but at students who want to become language or area specialists. That means that the classes are intensive (which means difficult to fit into a demanding sci-tech curriculum and not scheduled to avoid required courses in the student's main discipline. And especially in an

> ideographic script like Japanese,...the vocabulary learned in a class concentrating on literature or history is not the kind needed by an engineer. Beyond the introductory level, specially designed Japanese language courses geared to science and technology are a major need in the country."[12]

We hope that this conference will lead to a new awareness of the national needs in the field of technical Japanese translation, as well as to increased interest in the excellent research literature, textbooks, monographs, and handbooks in Japan. We also hope that this awareness will lead to increased interest in the teaching of Japanese and in the development of courses in technical Japanese. Perhaps we can also hope for increased funding by government agencies, foundations, and industry to do the research needed for enhancing our translation capability and for developing improved teaching materials.

ACKNOWLEDGMENTS

In connection with the preparation of our textbook, Comprehending Technical Japanese, we wish to acknowledge the enthusiasm and dedication of our colleague and friend Professor Nobuo Inoue, as well as the financial support of the Office of International Studies and Programs, the Engineering Experiment Station, the Graduate School, and the William F. Vilas Estate Trust Fund, all of the University of Wisconsin. We are also indebted to the Department of East Asian Languages and Literature for their continued help and encouragement in connection with the development of courses in Technical Japanese at the University of Wisconsin. In particular we wish to express our thanks to Professor Akira Miura, with whom we have had many pleasant and profitable interactions, and to students who have communicated to us their ideas concerning the process of learning technical Japanese.

FOOTNOTES

1. E.E. Daub, R.B. Bird, and N. Inoue, Comprehending Technical Japanese (Madison: University of Wisconsin Press and Tokyo: University of Tokyo Press, 1975).

2. According to a recent report, "The Reading Course in Scientific and Technical Japanese," given by the Centre for Japanese Studies at the University of Sheffield, requires no previous knowledge of the language and requires just seven weeks in residence. For a list of their books and other resources used in the course, see Technical Japanese Translation, no. 13, March 10, 1984, pp. 20-22.

3. J. Berndt and E. Yasui, "Japanische für Wissenschaftlicher und Techniker," (Leipzig: Veb Verlag, 1977).

4. O. Mizutani and N. Mizutani, An Introduction to Modern Japanese, (Tokyo: The Japan Times, Ltd., 1977).

5. E. H. Jorden (with the assistance of H.I. Chaplin), Beginning Japanese, parts 1 and 2 (New Haven: Yale University Press, 1963).

6. Y.M. McClain, Intermediate Japanese Reading Aids, vols. 1 and 2 (Tokyo: Hokuseido, 1971 and 1973).

7. "Let's Learn Japanese," Japan Broadcasting Corporation, Tokyo, 1975.

8. R.B. Bird, "Scientific Japanese: Part I, Kanji Distribution List for Elementary Physics; Part II, An Elementary Kanji List for Engineering Experiment Station," Report no. 33, 1967.

9. D.P. Crowley (with the assistance of Y. and Y. Kawata), Manual for Reading Japanese (Honolulu: University Press of Hawaii, 1972).

10. "Research on Vocabulary in Cultural Reviews," National Language Research Institute, Tokyo, 1957; "The Use of Written Forms in Japanese Cultural Reviews," National Language Research Institute, Tokyo, 1960.

11. Y. Kikuoka, Japanese Newspaper Compounds, (Rutland, VT: Tuttle, 1970).

12. R.B. Gillmor and R.J. Samuels, ed., "Japanese Scientific and Technical Information in the United States," Proceedings of a Workshop at MIT, January 1983, National Technical Information Service, p. 16.

COMMENTARY: A CAUTIOUS AND DELIBERATE STRATEGY IS NEEDED

Justin L. Bloom

I take a more cautious view toward the need for Japanese language training by American scientists and engineers than has been expressed by others at this conference. This view must be accepted in the context of my being strongly in favor of the study of Japanese by anyone, including the scientist or engineer, who is properly motivated and who understands what is involved in the undertaking. My rationale follows.

If the current flurry of interest in Japanese language training is based almost entirely on the fear of Japanese economic successes in high technology, this is not a sufficient motivation to sustain the student of the language through the many years of study that are required to achieve mastery. We have seen similar episodes in the past with German and Russian, but the problem is much worse with Japanese because of the greater difficulties involved in learning a language based on memorization of thousands of ideographs and their combinations. The student of Japanese should not be driven by external and perhaps temporal concerns about Japan, but by an intense personal interest that includes acquiring a knowledge of Japanese history, culture, and the intricacies of personal communication. He or she should be prepared to make a professional career of being involved with Japan or its people, to justify the investment of time and energy needed to acquire language proficiency alone, not to mention the requisite ancillary background noted above.

Putting aside the emotional aspects of training Americans to read the Japanese scientific literature, the jury appears to be out at present on the issue of whether there is an actual need for a greatly expanded flow of currently untranslated technical information from Japan. People who take this as a given often are the suppliers of information, but the users are relatively silent. It is possible -- until demonstrated otherwise -- that U.S. corporations and universities already have most of their needs met through a number of channels: (a) English-language technical journals already published in Japan; (b) English abstracts or indices of the Japanese-language literature provided by professional societies or commercial services; (c) in-house translations to meet individual corporate or academic needs, sometimes arranged through subsidiaries based in Japan and using native language speakers as translators; (d) use of bilingual consulting services. By unscientific, anecdotal methods, I have been able to ascertain that all of these forces are at play to some extent. A more legitimate survey conducted by the Conference Board in 1983, concerning acquisition of foreign technology in general, confirmed these impressions and did not indicate that there were problems in obtaining desired information.

There may also be more mundane reasons for the apparent silence of potential users. In some cases, American corporations or even academics do not appear to be interested in Japanese technical accomplishments at the scientific level. This may be due simply to the notorious Not-Invented-Here syndrome or it may be due to the burgeoning of the technical literature to the point where no individual appears to be able to keep up with all the developments in his or her field of specialization. In the latter case, there would be a natural tendency to follow indigenous developments first, and to search the foreign literature next only on a residual, time-available basis.

Another factor that should be considered is what appears to be a surplus of Americans who are fluent in Japanese and who are available to serve as translators or interpreters. They are having trouble finding jobs, although some of them are skilled in various technical fields. It would seem appropriate to absorb these people into the information acquisition system before undertaking the expense and time of training others. Related to this is another factor which I think will be demonstrated conclusively at a later time: a scientist who has learned the Japanese language is not going to be willing to spend his career working solely as a translator. He will spend as much of his time doing translations in support of his own work as required, but he will be most reluctant to perform this service for others. Thus his contribution to the pool of technical knowledge obtained from Japan will be variable and possibly limited. Of course, enough people in this category will have an impact, but not what I think is intended.

The last point that I wish to make is that it must be recognized that reading the foreign technical literature is only one device by which technical personnel learn from others. They also rely on personal contacts, attendance at conferences, visits to laboratories, and other similar interactions. One reason for this is that information in the published literature is often one or two years old before it is seen by others, and it may be incomplete. Therefore, anyone who wishes to be fully apprised about what is going on in Japan will have to spend some time there -- doing all of the things mentioned above. It would be helpful to be able to speak the language colloquially, and that is a challenge that adds another layer of difficulty to learning to read technical Japanese.

Question: How many young American scientists and engineers are prepared to make the kind of commitment I have described, and how many would be employed creatively for most of their careers once they have discharged their part of the implied bargain? The answer is unknown at this stage, and it is because of this double uncertainty that I suggest a cautious, deliberate approach.

II

GETTING AT JAPANESE TECHNICAL DATA: AN IMMEDIATE PROBLEM

ESTABLISHING A JAPANESE HIGH-TECHNOLOGY INFORMATION COMPANY IN THE UNITED STATES

Herman Baron and Tomoyuki Satoh

For the last twenty years, the primary flow of scientific and technical information has been from West to East, especially from the United States to Japan. This cutting-edge information has enabled Japanese manufacturers to challenge successfully American companies, and in most cases to establish significant market shares in industry after industry. Automobiles, steel-making, shipbuilding, plastics, home electronics, computers, ceramics, magnetics, textiles, and other American industries now recognize the full extent of Japanese competition.

Government and corporate efforts to monitor Japanese scientific and technical information have been scattered and ineffectual, with attention aimed mainly at establishing a "listening post" in Japan, translating a few articles, or subscribing to several journals. No government agency has taken the lead to monitor, on a large scale, the wealth of material openly available in Japan.

Xerox Corporation is launching a new information service, Japanese Technical Information Service (JTIS), which will bring to the West the top 1,000 journals published in Japan in the fields of high-technology and business. The journal articles will be abstracted into English and made available for about $5,000 as a monthly subscription service containing the 60,000 abstracts created each year. In 1986, the computer data base will be offered through a major online vendor.

The 1,000 journals were selected after a vigorous review of the Japanese journals covered by both Western abstracting and indexing services and the Japan Information Center of Science and Technology (JICST). Various worldwide periodical directories were checked, and recommendations were made by corporate, university and governmental librarians and researchers in the United States. The tentative journal list was then reviewed by Japanese research scientists, and their suggestions were incorporated into the final list. Emphasis was placed on the most advanced technological fields, those which Japan has targeted for future development: computers, microelectronics, robotics, biotechnology, ceramics, CAD/CAM, automotive technology, etc. Business journals will also be included and will cover management, finance, quality control, operations research, marketing, and so on.

A key feature of this service is high-quality abstracts (100-150 words in length) prepared by a select group of abstractors and translators who are fluent in both Japanese and English and have technical training. Half of them are American. This "electronic cottage network" will be reading, abstracting and translating material from Japanese into English.

It was not easy to locate this corps of high-level specialists, but it was not an impossible undertaking. We followed traditional avenues of recruiting employees, such as placing advertisements in specialized publications, and we also met prospective translators when we presented papers at information conferences. We also utilized our extensive network of contacts in the United States, Europe and Japan.

Networking is at the heart of the entire concept of JTIS. As an international information provider, we are keenly aware that we have obligations not only to American industry, but to the entire spectrum of researchers worldwide who look to the U.S. and Japan for leadership and breakthroughs in the coming years. English is the international language of science and technology, and JTIS hopes to make a major contribution by translating more foreign technical information into English. On the other hand, vast numbers of researchers in the United States, Europe and Asia are more comforable reading technical information in their own languages. The time is soon approaching when automated language processing will allow any researcher, in any part of the world, to scan and read in his or her own language from any other language.

When that time comes, JTIS will be there with the technical and marketing expertise to serve those researchers. In the meantime, our mission is to provide an open exchange of technical and business information between the West and the East. We invite your participation in this important undertaking.

THE PROBLEMS OF OPERATING A TECHNICAL TRANSLATION COMPANY IN JAPAN

Larry D. Brouhard

In the ten years I spent in the Air Force, in addition to learning that they don't always put you where you want to be (I joined as a trombone player and became an electronics technician!), I learned how important it is to write technical instructions and procedures that even 2nd lieutenants can understand.

My experience with technical translation and rewriting in Japan has been mainly in the telecommunications and computer fields. By describing the problems that one large computer and telecommunications systems manufacturer in Japan faces, and by outlining the problems I have had in providing translation and rewriting services to that company, you will have an idea of what the translation business in Tokyo is like. For selfish reasons, I'll concentrate on recruiting problems: I need technical translators and rewriters who are native speakers of English and who can ask, "How does it work?"

The quality of technical translations produced in Japan varies wildly, from random arrangements of English-like words to semi-intelligible finished documents in fairly good English. Ten years ago, a finished product from a typical company or agency doing technical translations might look like this:

> Casing shall be fully endured for the maximum belching pressure of operation in an operating temperature and for air pressure bearing test in a normal temperature.

Most translation agencies did not employ native speakers of English to edit, proof or rewrite. And translator capability ranged from that of the poor soul who did the passage above to translators who could produce "rewritable" English. Not many agencies were convinced that a check by a native speaker was needed. And because most customers didn't have native English-speaking staff members, few people had any idea of how bad the translations really were.

Japanese computer and telecommunications makers then began trying to sell systems abroad. It became necessary to export information as well as equipment. My good fortune resulted from the misfortune of one of these makers.

The company had sold an electronic switching system to the telephone authority of a Southeast Asian country. The first several thousand pages of documentation were shipped, but the customer sent it back with a message that essentially amounted to a blunt "Get serious!" I was lucky enough to get the job of rewriting some pages of the document.

That was in 1975. In 1977, the company formed a subsidiary to act as a clearing house for company translations. ICCS was selected to provide native speakers of English to check the quality of translations done by vendors and to rewrite translations done inhouse. That subsidiary now processes 1,000 to 15,000 200-word pages a month -- all Japanese to English translation -- and ICCS has ten people permanently assigned there.

A "gaijin" is a "gaijin"

Our first job as checkers was to convince the vendor agencies that we wouldn't accept translations that had not been checked by a native speaker of English, but non-compliance was the rule. In the earlier days, about half of the translations we checked were about the quality of the passage above. Another 25 percent were rewritable, but had not been checked. The other 25 percent had been checked. We rejected more translations than we passed in those days, so the agencies began looking for "gaijin" (foreigners).

For you to understand the recruiting problems, I need to say a little about the English conversation business. In Tokyo, English conversation is the mizu shobai (honky-tonk life) of the language services industry. There are some excellent, well-qualified people in the business. To find rewriters, agencies ran ads in the Japan Times, a Tokyo-based, English-language daily; foreigners tired of teaching English conversation answered them.1 Just as all Westerners are not English teachers, neither are they technical rewriters. There were many problems. My requirements for an ideal rewriter applicant include:

- Technical background in electronics, computers, or telecommunications
- Ability to speak and read Japanese
- Good writing skills
- Minimum 18-month commitment
- Presentable appearance
- Willingness to work for low pay

Recruiting was easier in the early days because there was a sizable pool of U.S. military veterans from whom to choose. Many had been technicians in the armed services and had attended Sophia University's international division or International Christian University. They could speak Japanese, were going to be in Japan for a while, knew the importance of "understandable" technical manuals, and were accustomed to low pay.

By about 1979, the pool had dried up. There was still plenty of response to our classified ads in the newspaper, but of 30 applicants, there would often not even be one who had more than a couple of the requirements listed above. I began leaning more on Japanese ability than on technical background, and that was a mistake.

The "Jimmy" and the "Temple Dweller"

A "Jimmy" is a Japanese who has mastered English and little else. He discovers on his return to Japan after graduating from an American college or university that he is qualified to do little other than teach English. But he doesn't even have much chance of that, as everybody wants "native" speakers. He's the guy with the name card that has a Western nickname bracketed by double quotes. Foreign firms hired him because he could speak to them, but then found that he couldn't speak to the Japanese.

The "temple dweller" is a Western foreigner who has mastered Japanese and little else. He comes to find the "real" Japan, to visit temples and shrines, perhaps to get beat up on the weekends at a Zen temple. The temple dweller doesn't do us much good in the technical translations and rewriting field. It is not just that he doesn't have the technical background to ask "how it works," he usually is not even very interested in how it works.

The Language Major Outside the Classroom

Is there life outside the classroom for a language major? To keep from getting into trouble, I'll be timid and say only that the prospects seem limited. Just as with Jimmy, who spent all those years mastering English and little else, the same may be true for the American who specializes in Japanese, judging from what I've heard from Japanese area specialists whose services no one wants. The goal should not be the language itself, but the communication accomplished using the language.

Please do not cut back on the number of persons majoring in Japanese. Instead, broaden the base of Japanese studies by encouraging majors in science, math, and other technical subjects to study Japanese in addition to their specialty.

In 1980, I was looking for systems software specialists who also knew Japanese. Of the 30 or so persons who answered the ads in the San Francisco, Santa Clara, Los Angeles and San Diego papers, only one knew any Japanese. He would have been perfect for the job, but his wife, who was Japanese, decided she didn't want to return to Japan. The person I hired didn't know any Japanese, but was an excellent systems person and a good writer. He has worked out so well that he is managing our native-English checkers during the day and studying Japanese at night.

Again near the end of 1982, I ran ads in the above cities, adding Phoenix to the list. This time nearly 150 persons responded. I was looking for three entry-level technical writers. Almost all of those who responded were out of work. Many were victims of the recession; many others would be employed only in the best of times. Not one of the two dozen or so who were interviewed knew any Japanese. One came to Japan, worked three-fourths of his contract, then left for another job. He, too, did excellent work but had a hard time adapting to life in a Japanese company.

Recently a man called from California looking for an American chip designer who could speak Japanese. He wanted the person to act as a liaison and interpreter during a three-month orientation that the American company was holding for its Japanese research staff. The man said he had talked to about 40 Japanese-to-English translators in the United States without finding anyone who knew anything about chip design. He was pretty discouraged by the time he got to me.

What the Japanese Are Doing

Computer manuals in the United States are horrible, but in Japan they are much worse. Some blame the Japanese language itself, saying that it is a vague language that doesn't lend itself to specific logical presentation.

Until just a few years ago, there was no technical writing field in Japan, nor was there much of a commitment to producing good manuals. Someone said you had to have manuals, so companies wrote manuals. That they were not understandable didn't really matter; after all, nobody was complaining.

One company discovered that its original manuals, even if reasonably well-translated, were not usable. Someone _had_ complained. An aggressive manager within the company convinced them to go a step past translation, and send the translated manuals to the United States to be rewritten again by experts in whatever specialty the manual covered. This approach has been moderately successful. At the same time, the company has been trying to upgrade the quality of its original Japanese manuals by giving technical writing courses, and by trying to convince managers to emphasize good writing skills and document engineering.

The company is not alone in its efforts. The Japan Society for Technical Communications (JSTC), which is part of the Japan Management Association, started administering 4-level technical English proficiency tests four years ago. The growth in the number of persons taking the tests was only moderate until last year, when it showed a 68 percent increase. More people are also attending the JSTC's seminars on technical writing. Japanese manuals are bound to get better.

Machine Translation and More Jobs

Machine translation will not do away with the need for Japanese language skills. On the contrary, it will increase the need for them.

A basic tenet of translation is that the translator be a native speaker of the target language. In Japan, that is impossible. If just one company in one field produces 15,000 pages per month (3 million words), think of the volume throughout Japan. Just to satisfy the needs of the one company would require fifty native-English translators producing 3,000 words a day for twenty days a month, twelve months of the year.

Right now our emphasis is on rewriting the translations of non-native English-speaking translators, who all tend to make similar mistakes. A rewriter who understands the subject and reads just a little Japanese can muddle through.

When machine translation becomes "rewritable," which I am confident it will, the need for technical rewriters will increase sharply, especially in the early stages. But this new batch of rewriters will have to know more Japanese than their predecessors. And they will also have to have some idea of how machine translation works so they can guess why the computer pulls the boners it does. They will need to be able to read between the lines of the Japanese originals.

The Age of Information

World's first, world's best, world's most, world's highest reliability, epoch-making, and the age of information -- these are all buzz words in Japan. There are times when I wonder if anything is being done anywhere else in the world. Hyperbole aside, no talk related to Japan would be complete without mentioning "the age of information."

Until just a few years ago, Japan seemed interested only in importing information and exporting products. Indeed, the Japanese attitude seemed to be that foreigners couldn't possibly understand the Japanese anyway, so why bother. On the American side, there seemed to be an attitude of technical arrogance. "The Japanese merely copy things. They never really come up with anything new." That attitude still prevails among many computer software developers in the United States.

The turning point, on both sides, seemed to be when Japan overtook Detroit as the world's leading car producer. U.S. companies became interested in Japanese management and production techniques. The Japanese, on the other hand, seemed to become less timid. As a 19-year-old

Air Force two-striper in Japan, the attitude toward me was: "You're an American. You must know everything." Now I'm 40 and it is: "You seem to know a lot for an American."

And we don't hear very much these days about the inscrutable Japanese from the Japanese themselves. In fact, they are taking the export of information very seriously. According to the September 30, 1984 issue of the Japan Times and the January 5, 1985 Asahi Shimbun, Japan aims to receive at least 50,000 foreign students in various areas of study in Japanese universities by the year 1990, and 100,000 by the turn of the century. The level right now is just over 8,000. Crash programs in the teaching of Japanese as a foreign language are being developed. There are presently about 1,000 teachers of Japanese for foreigners in Japan. By the year 2,000 they will need as many as 10,600.

What kind of information will the Japanese be exporting? Surely students from lesser developed countries are going to be more interested in low-, medium- and high-tech subjects than in the aesthetic placement of rocks in the garden at the Ryoanji, a famous temple in Kyoto.

The age of information also brings with it the need for every person in a country like the United States to be involved somehow with technology -- the home computer, videotex systems, telephones that let you transfer money between bank accounts at the touch of a few buttons. These things are here already.

The Japanese lead the world in the production of automobiles, televisions, video tape recorders, and computer peripherals, and they are doing exciting research with semiconductor devices. It is popular to blame some of Japan's success on generous government subsidies, restrictive government import policies, unlimited access to the financial resources of a chosen bank, company-sponsored unions, and even pervasive sexism. Some also like to insist that they "borrowed" all of ther basic technology. I don't know. But in Japan a lot of people get to work very early and stay very late.

At any rate, the future is exciting. Much is to be gained by technical cooperation between U.S. and Japanese companies. U.S. companies have technical expertise that Japanese companies can use. Likewise, Japanese companies have a lot to offer. But right now, the transfer of information is lopsided.

The Japanese gather information in English, sift through it, then translate what they want into Japanese. U.S. companies seem to have to make do with government white papers or whatever information they get from the trading companies they work through.

A Plea for the Future

I don't have a plan to fix the world. My goal is a selfish one. I want a larger pool of technical persons who know Japanese to choose from. I would like to see Japanese language and technical/business communications courses offered to students in all areas of study.

If, indeed, 100,000 foreign students are to study Japanese in Japan by the turn of the century, now might be a good time for U.S. colleges and universities to start pressing for aggressive student exchange programs. Another area to look into might be that of working visas in both countries. Japan and Australia presently have a "working holiday" arrangement whereby persons can work in either country for a year without need of sponsorship. The working holiday arrangement would be ideal for a prospective translator or engineer to get everyday experience in Japanese. This experience is essential.

This next suggestion may sound ridiculous, but why not promote the Japanese board game of "go" among computer science majors. In the past nine years, I have known at least ten rewriters and translators who had majored in a computer-related field and who came to Japan for the sole purpose of improving their "go" abilities. I am not a "go" player myself, but computer people seem to go for it in a big way.

Conclusion

Future translations of technical materials from Japanese into English will surely be done by computer. But they will still require extensive rewriting by editors who can read Japanese. Maybe it wouldn't be outlandish to add a couple of new areas to technical writing courses, those of Japanese translation and editing of machine translations.

I do not expect all rocket scientists, computer specialists, or chemists to study Japanese. And people with far more experience in the translating business than I have continue to argue the pros and cons of the generalist versus the specialist translator. Regardless of which side one takes, however, the need is, and will continue to be, for people who can ask: "How does it work?"

THE JAPAN INFORMATION CENTER OF SCIENCE AND TECHNOLOGY

Toshihiko Kanda

The Japan Information Center of Science and Technology (JICST) is the central organization of information activities for the advancement of science and technology. The Center was established as a special nonprofit organization under legislative act on August 16, 1957. JICST is under the executive control of the Science and Technology Agency, Prime Minister's Office.

JICST is financed by income from two sources: governmental support and subscription and service fees. In principle, the cost of collection and processing is provided by the government. Service costs are met from sales income.

JICST implements the following activities in order to encourage scientific and technological information works in Japan:

1) collecting scientific and technological information comprehensively on a worldwide scale;

2) processing this information systematically;

3) disseminating information rapidly and appropriately to organizations and individuals, regularly or upon request;

4) offering assistance to other far-reaching services to encourage their scientific and technological information works.

To these ends, JICST engages in the following services:

Online information retrieval service
Publications of abstracts from journals in science and technology
Photo duplication service
Translation service
Manual search service
SDI and RS services
Other services

JICST Online Information System (JOIS)

The development of JICST's Online Information System could be summarized as follows:

JICST started the experimental online information retrieval service in April 1976. Full-scale operation of JOIS began in June 1978, utilizing the public telecommunication lines. In October 1980, the inter-connecting network was constructed with the leased and public telecommunication lines in ten major cities of Japan, thereby establishing a public telecommunication line network that eliminated regional differences in access and utilization capability.

A more advanced online system, JOIS II, has been in operation since April 1981. JOIS II is an international commercial online system holding 2,600 passwords and is at present accepting 70,000 sets of questions per month. The number of JOIS serviced data bases has increased to 17, and system capabilities have been expanded every year.

The main functions of JOIS are:

1) Search capabilities: The system performs the fullmatching as well as right and left hand truncation on the basis of key words that are entered by the terminal. It is also capable of logical search of articles by using "AND," "OR," and "NOT" functions, as well as limiting the number of searches by such means as language, author, publication date, etc.

2) Auxiliary search capabilities: The system can display the following or preceding keywords in the Japanese alphabetical order of the keyword that is entered.

3) Registration of search command: The search command may be prestored or modified.

4) Capability to assist conversation: The system provides announcements from JICST, data base guide, request for online orders for copies, etc.

5) Functions to be used by personnel controlling the terminals: The system performs registration or renewal of passwords, displays charges for the use of data bases (including the total of accumulated charges), etc.

The following chart shows the growth of utilization of JOIS in Japan:

	1977	1979	1981	1984 (Sept.)
Number of users	20	500	1,100	2,600
Number of hours used per month	200	1,200	2,400	5,900

From this, you can see that the growth in recent years exceeds 30 percent per year.

Characteristics of JOIS

A user can use the system to search freely a number of the JICST data bases including "JICST File on Science and Technology," containing scientific and technological documents of high quality published in more than 50 countries, or the "JICST File on Medical Science in Japan," a data base of important medical documents available in Japan, all of which have a worldwide reputation. In addition, users can have online access to the "JICST Thesaurus File" and "JICST Holding List File" which are used as supporting files to search the data bases noted above.

The user may also utilize the "JICST File on Current Science and Technology Research in Japan," the summary of research activities currently in progress at the major public research organizations in Japan, to know immediately who is doing what kind of research and where in Japan.

In October 1984, JOIS started the "Nikkan Dogyo File on New Technology and Products in Japan," and the "JICST File on Government Reports in Japan" for domestic as well as for future overseas customers.

An advantage of this system is that it can store in advance all search commands on a specific subject so that the system can conduct an automatic search on monthly updated data of each file and deliver the results to the users.

JICST is at the same time a data base producer and distributor, and holds all of the original full texts of the abstracts derived as the result of the JOIS search. As it is, our users need only to enter the document number of the article and the number of the abstract from the terminal when ordering.

Operation Hours and Charges for Use of JOIS

Operation hours for JOIS are from 9 a.m. to 8 p.m. Japan Standard Time. The charge for the use of JOIS data bases is 12,000 yen (approximately $47) per hour for connect charges and 10 yen (approximately three cents) per each online hit charge, except for the "Nikkan Kogyo File on New Technology and Products in Japan," which is priced at 14,000 yen (approximately $56) per hour for connect charges and 10 yen for each online hit charge. The user must bear, in addition, the telecommunication charge and the agent commission. JICST also offers such privileges as the "Connect Time Charge Discount" and "Discount of New JOIS Users" for those users who may want to take this advantage.

JOIS Contract and Payment of User Charges

Users who wish to subscribe for the JOIS services are requested to contact the JOIS sales agent in their country. JICST has entered into the JOIS marketing agreement with these agents. These agents are authorized to conclude the JOIS Service agreement with each end user. The user is asked to complete the form for "Subscription for JOIS Service," which should be sent to JICST through the contracting agent. Payment for the user's service charges is also to be made to the agent on a monthly basis. There is no initial charge or payment.

Future Prospects of JOIS

JICST produces and offers data bases covering every field, particularly all the latest information on Japanese scientific and technological developments. The development of science and technology has brought rapid advances in Japan so far. In the field of electronics, new materials, medical science and biotechnology information, Japan is considered one of the most advanced information organizations in the world. Up to now, JICST's main role has been to collect information related to the development of science and technology in order to provide Japanese researchers and scientists with the latest and most important information in these areas.

For the future, however, JICST considers as its most important task development of a method of disseminating in English to the whole world the latest information on Japan's science and technology.

For the time being, at least, JOIS will have to provide information in the Japanese language, and this requires the use of an English/*kana* type or a *kanji* terminal. However, beginning in April 1986, JICST will inaugurate a service in which searching can be made by ordinary ASCII terminals, since the titles and key words for the JICST files of Japanese documents will be translated into English by then.

In addition, JICST is planning to establish adequate infra-structures to enable large-scale translation and copying service for its Japanese documents in response to requests from overseas. Also, along with the document search, the JOIS service will be expanded to include a factual data base on chemical compounds.

III

THE FAILURE OF U.S. NATIONAL LANGUAGE POLICIES

A NATIONAL AGENDA FOR JAPANESE LANGUAGE INSTRUCTION

Richard D. Lambert

While I am a bit of an outsider in Japanese studies, my proper guild being South Asian studies, I am involved in a number of national level activities that are relevant to your deliberations. In particular, I would like to talk about a few national developments with respect to foreign language instruction, with particular reference to Japanese.

First, let me note that there has been a sudden surge of interest in foreign language instruction, not so much by students -- although enrollments in Japanese have increased immensely and language enrollments in general are on their way up again -- but among those concerned with educational policy and with national policy more generally. There has never been a time in my memory when there has been a more open moment, a more encouraging climate for the development of a fresh American national policy with respect to the teaching of foreign languages. Many people in Congress, throughout the federal government, in the private foundations, and among the general public have become convinced that the foreign language competencies of both Americans in general and among those in positions where knowledge of a foreign language is essential to job performance are far too low. Let me just mention a few very recent reports that indicate this interest.

If you have not already seen them, you should look at Paul Simon's eloquent broadside, The Tongue-Tied American, and the report by the Congressionally mandated National Advisory Board for the U.S. Department of Education's International Programs. The latter is entitled Critical Issues in International Education: Recommendations for Action; it can be obtained upon request from the office of the Secretary of Education. For a national policy statement of special relevance to Japanese and similar non-Western languages, you should see the Association of American Universities' recent report, Beyond Growth: The Next Stage in Language and Area Studies. It is available upon request from the AAU. The language chapter in that report is of particular importance for the future of federal and foundation support for Japanese language instruction, and you will be relieved to know that Eleanor Jorden had a major hand in writing it.

In part as an outgrowth of these reports, there have been some promising developments in legislation and federal agency policy with respect to support of efforts to improve foreign language instruction. For instance, in the recently passed math and science bill, a fund of several million dollars for this fiscal year was tacked on, to be allocated by the Secretary of Education for the improvement of instruction in "critical languages." The list of critical languages has recently been published in the Federal Register and it includes, of course, Japanese. The National Institute of Education has just closed a competition to allocate some $7 million for the establishment of a center for research and experimentation in language instruction. While most of this will go to bilingual education, about a third will be utilized for the improvement of second language instruction, including Japanese. The Department of Defense has convened an intra- and inter-agency committee to develop an implementation strategy for the recommendations in Beyond Growth; their principal interest is in the recommendations of that report with regard to language instruction.

In addition, Title VI of the Higher Education Act, which contains the most important funding for research in and teaching of the less commonly taught languages, including Japanese, is up for congressional reauthorization. I understand that preliminary congressional hearings will be held in May of this year. In the meantime, a sub-committee of the National Advisory Board of the Department of Education is preparing a report advising the Secretary on what the Department's position will be on the future language of that act. Public testimony was solicited in writing and in public sessions held by the sub-committee in Washington, San Francisco, and Chicago. I am chairing that subcommittee and I can report that while the national umbrella language associations, such as the Modern Language Association, the Joint National Council on Language, and the Russian and German language teachers' associations gave us their views, I was somewhat disappointed that the Asian language teachers' associations, whose members are most affected by Title VI, did not offer testimony. This is particularly regrettable since some of the most important sub-committee suggestions for change in the legislative content and programmatic implementation of Title VI will relate specifically to on-campus instruction in the less commonly taught languages, including Japanese.

All of this is just part of a general churning around, much of it constructive, that is taking place with respect to foreign language instruction in the United States today.

I would like to take my remaining time this morning to discuss a number of issues that have emerged from this renewed concern with foreign language instruction, issues that are directly relevant to our effort to facilitate the flow of Japanese technology to the United States by increasing the pool of Americans in key positions who have a working knowledge of the Japanese language.

In the brief time allotted to me I can only make mention of six issues which I will refer to as the determination of instructional effectiveness, uniform measurement, scale, skill level, organization, and adult reference.

Determination of Instructional Effectiveness

For some twenty years, I have been an observer of the various regional guilds in language and area studies. My overall impression is that they vary considerably in the effectiveness of their language instruction on a continuum roughly paralleling the essentiality of the language in conducting research on the area, the difficulty Americans experience in learning the language, and the development of a self-conscious tradition of language pedagogy. Japanese instruction heads the list in general effectiveness, followed by Arabic, Chinese and Russian in about that order. The South Asian and African language teaching traditions are toward the bottom of that continuum. This overall general impression was reinforced just a fortnight or so ago when I observed students performing on an oral interview test midway through their training in the Tokyo Inter-University Center. I know no Japanese, so I cannot attest to the accuracy of their performances, but I was impressed by their ability to carry on a sustained conversation and to produce a sustained dialogue. Having been chairman of the University of Pennsylvania's South Asia program for 15 years and for several years Director of the American Institute of Indian Studies in India, I can assure you that that would not have been possible for those training to be India specialists at an equivalent stage in their careers, and this in spite of the much greater learning difficulty of Americans studying the East Asian languages compared with those of South Asia.

But we really have no idea whether the cluster of teaching technologies used to bring the students up to that point was the most effective set that might be used. We do not know because we have a surprisingly unempirical tradition in shaping language instruction. We seem to go in for wild enthusiasms -- the Lazanov Method, the Silent Way, Rassias, Total Physical Response, Suggestopedia, the Monitor Method, etc. -- each one establishing its supremacy by assertion rather than proof. Fortunately, these language teaching fads have not been visited upon Japanese language instruction, but even in Japanese we depend on teachers' intuition -- a measure of variable validity at best. We rarely think to collect systematic evidence of what works best on what kind of students at what level and for what circumstances. Instead of carefully controlled experiments that might begin to answer some of these questions, we have assertions as to why this or that technology of textbook must be better than the others.

Here we are again at this conference discussing how we might best tailor Japanese language instruction to the needs of engineers and scientists. But we do not build in from the outset any evaluation design that will enable us to sort out five years from now just what worked well and what was an utter waste of time. My own view is that the amount of classroom and extramural time we demand of our students to learn Japanese makes it incumbent on us to determine, with as much precision as possible, what makes the most effective use of that classroom time. However, as a profession, we tend to leave the answer to that question to a custom, repeating the way we were taught, or to intuition, our seat-of-the-pants hunches as to what might, perhaps, maybe, might work better. It is high time a self-conscious, highly empirical and experimental tradition is developed in Japanese language pedagogy.

One reason why we have so little tradition for measuring the relative worth of various pedagogical strategies is that we have no adequate measures of how much language competency someone has. All we usually know is the number of semesters or years someone has studied and perhaps a score or two on achievement tests geared to the materials actually taught in the classroom. This will no longer suffice, particularly if we are going to produce and market people who will be certified for their jobs -- especially non-academic jobs -- as having a certain amount of language competency. We need to share a common metric that indicates a general competency level in terms of real life, usable skills the person possesses, rather than how much of the textbook and classroom materials he has mastered.

Probably the strongest movement in the field of foreign language pedagogy today is the drive to make universal the adoption of a set of standard definitions of language competency derived from the oral interview of the Foreign Service Institute. As this testing technology is being extended to languages such as Japanese -- particularly those with difficult-to-master orthographies -- and as it has to deal with students at the lowest skill levels at which most academics are found, major problems of adaptation are being encountered. We all have a stake in the solution of these problems, since the application to Japanese of a metric geared to real-life proficiency, one shared with other languages, has a very high order of priority on our national agenda. For one thing, one of the recommendatins now being considered in the Title VI reauthorization is the creation of two tiers of FLAS fellowships. The first tier would be for beginning students, while the second tier would have an entry qualification of a relatively high level of demonstrated language proficiency. But to implement such a system, where are the tests that can be used to administer such a program?

The problem of uniformity in our proficiency measures will become even more urgent as we extend the scale of our operation to include many more and different kinds of students. Indeed, the whole organization of Japanese language teaching -- currently a kind of cottage industry -- will have to change if the numbers and types of students expand. To give you an idea of the ultimate in possible expansion, the Chief State School Officer for the state of Utah recently told me that he had been offered free access to a satellite and he was proposing to use it to beam Japanese language lessons into every secondary school in

Utah, plus possibly Nevada and Texas as well. He reported that he had had a promise of enthusiastic assistance from the Japanese government.

Now that is a real change of scale. Our current cottage industry will burst if it tries to service such a venture. Where are the teachers, where are the text materials, where are the strategies for teaching so many students dispersed over so wide an area? I have been impressing on that school officer the absolute necessity of monitoring his experiment, but where are the normed tests he can use to see what worked and what did not work, and where are the procedures for mounting a convincing evaluation study?

Another issue that the expansion of Japanese language training to scientists and engineers will raise is the question of the level of language skill we teach. At least some students who expect to use their language skills will need a truly high -- that is, near-native -- level of competency. However, by and large our current teaching system is aimed at the lower skill levels. I am watching with considerable interest a proposed experiment linking several American universities with one on mainland China to teach Chinese to Americans at a very high level of competency; for example, to take up where the Taiwan center tends to leave off. Similar attention to upper level skills is called for in Japanese as well.

Problems of scale, a focus on higher levels of skills, as well as the shifting character of our clientele will dictate changes in the organizational pattern of our delivery system for language instruction. There are already strains in the system, even in the current academic setting. The current allocation of three or four hours of classroom time a week spread over perhaps two years is neither long enough nor intensive enough for maximum effectiveness. It just takes longer than we normally have in classroom time to get a student to a genuinely usable level of competency in Japanese. The situation is better for students who can take a year's language training in Japan, but most students cannot.

Beyond the limitations on time, our teaching system is geared almost exclusively to the needs of students registered for credit in a semester or year-long course. This procrustean bed just does not fit many of our potential clientele whose demand for language instruction comes later in life and for whom the pieces of time they can give do not come in neat, semester-long chunks. We can say we will not meet their needs and leave them to the abominable fluency produced by the quickie proprietary schools. I would suggest, however, that the field has both an opportunity and a responsibility to provide facilities for teaching adults outside the normal classroom format and rhythm.

Speaking of these non-student clienteles reminds me of another general agenda item. The Japanese language profession has an opportunity to lead in what I consider a necessary transformation of our perspective on national foreign language policy. At present, we concentrate all of our attention on the teaching delivery system for language education and the students enrolled in it. A primary focus, however, should be on the language competencies of adults. For instance, it has always surprised me that we do not have the kinds of national needs or use surveys that are so common in Europe. These are surveys of the actual use of foreign languages either by the general adult population, or by particular occupations. Such surveys tend to anchor national language policy in the societal use of foreign languages, either existing or desired. For most languages, we seem to pay no attention to adult utilization, preferring to fight the battle of language requirements for college students with no idea as to whether they will actually use the language or not. For Japanese, however, the current boom in adult demand might enable it to tie the training process to adult use as well as early training of school-age students.

Looking at language competency from the adult perspective also calls attention to the problem of language maintenance and language loss. We put all of our resources into first-time language acquisition, none of it into making sure that the language skills are retained. If we are really serious about providing language competencies to scientists and technologists, then some attention must be paid to helping individuals to sustain those competencies over periods of intermittent use.

Let me close by saying that I have been able to mention only a few of the issues of language instructional policy which face the field of Japanese studies as it expands to include science and technology. I have discussed issues of empiricism, a common metric, competency level, organizational style and skill maintenance. There are obviously others. There is a major national dialogue in progress concerned with the future of international studies. The language teaching community must get its act together, decide what it needs to do to meet the new demands placed on it, and organize to get its agenda front and center. The Japanese language teaching fraternity is in an excellent position to help set the national agenda.

NATIONAL LANGUAGE POLICIES AND THE 99TH CONGRESS

J. David Edwards

One might conclude that, if we were allotted 15 minutes to discuss "National Foreign Language Policies in this Congress," we might be left with a considerable amount of time on our hands. One might also conclude that in assessing a prognosis for the 99th Congress, you need a crystal ball that registers only the single word -- deficit. The latter is probably correct. The former is not. This Congress will have to address some issues of significant concern to foreign languages and international education. Before addressing current and future legislation, allow me to take a moment to sketch the current policy context concerning languages and to describe just a few of the policies accomplished in the 98th Congress.

The current public policy context can be characterized as "guardedly optimistic." Some exciting things are happening with regard to education in general and language studies in particular. The last four years have produced increased media attention, growing public support, and most importantly, professional unity and organization. We have been able, with a lot of help from our friends, such as Paul Simon, to translate these factors into improved and new policies, increased legislation, and major policy changes in the states.

First, there has been a general revival in language study. Enrollments are up considerably. For example, the State of Virginia has more students studying foreign languages now than at any time since World War II. The Modern Language Association reports that higher education enrollments have increased in all languages except Hebrew, Portuguese, and the Classics. Chinese and Japanese continue to grow like Topsy, and Russian, German, French, Italian and Spanish have made notable gains. Last year seventy colleges and universities reinstituted language requirements for admission or graduation (and that will have a trickle-down effect).

Second, the language profession has moved to encourage communicative competence and develop proficiency-based standards. ACTFL and ETS have already put in place a series of oral proficiency tests based on a revised version of the FSI scale for the commonly taught languages. These tests allow us to tell students with some exactitude what we can and cannot do and what they have or have not done.

Third, according to the former Secretary of Education, 165 state-level task forces have been created to examine education in all fifty states. Most of these task forces address foreign languages. Eleven states have already introduced legislation to upgrade language requirements. In Florida, for example, the state is providing incentive funds for foreign languages in the elementary schools, awards for outstanding high school students, student performance standards, and requirements for admission to the state universities. New York is leading the way in adopting an action plan for global education. State aid will be available to school districts based on foreign language enrollments; all candidates for a Regents' Diploma must pass a proficiency examination; and by 1992, all students must have studied a foreign language by grade 9. To some, this suggests that the "New Federalism" is working. And it may be. The danger we must guard against in this Congress is that reduced federal funding and support would make these state programs more image than substance.

Nationally, the stage has been set for improved language study and facilities by over three dozen national commission and task force reports. These studies range from <u>A Nation at Risk</u>, the report of the National Commission on Excellence in Education, to more specific considerations, such as <u>Critical Needs in International Education</u>, the report of the National Advisory Board on International Education, which puts forth 19 specific recommendations for improving the quality of foreign language and international studies in the United States, and the most recent study, <u>Beyond Growth: The Next Stage in Language and Area Studies</u>, conducted by the American Association of American Universities for the Department of Defense.

From a policy perspective, there can be little doubt that these reports, the state activities, and the increase in public attention have assisted in creating among policymakers a positive mood and a genuine sense of urgency toward strengthening education and language study. This prompted a number of very important legislative initiatives in the 98th Congress.

Last year, an election year, the appropriation for education was $17.6 billion, a $2 billion increase over FY 84. For the second year in a row, international education and foreign language programs were zero funded in the budget request and, for the second year in a row, Congress responded with an increase to $32.05 million for this fiscal year. Exchanges received an increase of $35 million and two new programs -- the Soviet and East European Research and Training program and the Congress-Bundestag Exchange program -- were funded at $4.8 million and $2.5 million respectively.

In addition to increased funding, some very significant new programs also came out of the 98th Congress. The Library Services and

Construction Act amendments included a new Title V providing grants of up to $15,000 to state and local public libraries to purchase foreign language materials.

Also, the Department of Defense authorization bill created a new U.S. Institute of Peace, actually funded at $4 million, to provide fellowships to individuals and grants to universities and research groups to promote the study of international diplomacy and conflict resolution.

Finally, Public Law 98-377, the Education for Economic Security Act, was signed by the President on August 11. The final statute looks very different from the small, $80 million math/science bill that started out in the House initially. The law now contains eight titles, including one on equal access for religious groups, and one on asbestos hazard abatement. While Title VI, Excellence in Education programs, and Title VII, Magnet School Assistance, are important to us, the heart of the bill is Title II, which provides assistance for teacher training and instruction in math, science, computer training and foreign languages. The continuing resolution actually contained $100 million for this Title for this year, and the Education Department is currently moving to create the grant procedures to have the money in the field by June.

Languages are a stepchild in this legislation, but that's OK; originally they weren't in at all. Essentially, at the state and local levels, foreign language instruction and computer learning get what's left over after the schools have trained sufficient math and science teachers. At the post-secondary level, "critical" foreign languages compete with math and science to create cooperative programs for improving student performance. Finally, under the Secretary's discretionary programs, there are specific critical foreign language grants and grants for national significant projects.

Of equal importance is what did not happen in the 98th Congress. A new bill to provide scholarships for undergraduates from developing countries was introduced at the 11th hour. It had support from both the majority leader and the minority leader in the House and bipartisan report in the Senate, so it almost slipped through. In the end, its short life was extinguished in a conference committee.

The American Defense Education Act (son of NDEA), with its very strong provisions for foreign language education, came to the floor of the House in the final days. When it began to attract every election year issue from school prayer to busing to animal safety, it was tabled by its friends rather than risk its future.

As many of you know, the Foreign Language for National Security Act, providing $50 million for languages at all education levels, passed the House last February with a better than two to one margin. Despite our best efforts, this bill languished in the Senate sub-committee for seven months and finally gave up the ghost.

I mention these three bills because some version of all of them will come up again in the 99th Congress, and because they are all bills to which the language community is seriously committed. The caveat is, while they will most likely be re-introduced in the 99th Congress, any new spending bill, even one as modest as the foreign language bill, is going to encounter very rough sledding.

The mood of the 99th Congress is unquestionably one of fiscal restraint. Financial issues such as the deficit, defense spending and tax simplification are likely to dominate the entire life of this Congress. Illustrative of this, the Administration's FY 86 budget officially went to Congress on Monday, February 4. We identified 34 areas of concern to languages and international education in this budget.

The premier programs of concern to languages, Title VI, International Education and Foreign Language Studies, has once again been zero funded, as have the fund for the improvement of post-secondary education, library services and the U.S. Institute of Peace. My understanding is that Congressman Panetta and Senator Dodd intend to introduce a joint resolution similar to one they introduced last year, indicating that it is the sense of the Congress that Foreign Language and International Studies funding should not be reduced but should, in fact, be increased. For Title VI, at least, the two previous attempts at eliminating that program did actually produce slight funding increases. FIPSE also survived one attempt at elimination. The fate of the U.S. Institute of Peace, however, should be very much in doubt. This is not to suggest that any of these programs are secure. A 14-seat shift in the House favoring the Administration, the loss of 23 Representatives (Republicans and Democrats) who voted for the Foreign Language Bill last year, and a general mood that the budget must be reduced, all make the future of any of these programs uncertain.

Of particular interest to this audience is that the U.S.-Japan Friendship Commission and the Asia Foundation received funding proposals similar to last year's actual appropriations of $1.6 and $9.6 million respectively. On the other hand, the National Endowment for the Humanities, a major supporter of language research, would be cut by $13 million.

On the positive side of the ledger, the Administration has requested continued funding at $100 million for FY 86 for Title II of the Education for Economic Security Act, although Excellence in Education and Magnet School Assistance will be eliminated. Exchanges are to receive a $28 million increase. Humphrey Fellows, private sector programs and the Congress-Bundestag Exchange program all receive slight increases. This is consistent with the Pell Amendment passed in 1982 insisting that exchanges be doubled over the next four years. As we approach the end of the funding cycle in 1986, we can anticipate a leveling off in funding for exchanges in the second year of the 99th Congress.

We can also anticipate that the budget debate will pretty well occupy Congress's attention for the next few months. After that, they may begin to seriously address reauthorization of the Higher Education Act, which expires this year, although there is an automatic extension through next year. Consequently, Title VI, International Education and Foreign Languages, along with the rest of the Higher Education Act, must be addressed in the 99th Congress, and I would not be surprised to see this as the only major education issue addressed by the 99th Congress.

A number of Congressmen have begun to draft their proposals for Title VI; Professor Lambert's sub-committee of the National Advisory Board has held hearings and is drafting its proposals; a task force exists in the higher education community that has met a number of times and has its suggestions pretty well in place; and the language profession has offered its suggestions to the Lambert sub-committee, interested legislators, and anyone else who will listen. My sense is that we have our act pretty well together on Title VI and, while we may have some disagreements, we are not so discordant as to have Congress decide "there they go again" and proceed to make decisions for us.

With regard to the total reauthorization package, Title V, Teacher Training, is quite important as well. Given the language profession's concern with teacher shortages, proficiency standards, and some real questions about quality raised by the shortages, we will be very concerned, in any new provisions for teacher training, to include options for quality pre-service and in-service language programs.

In conclusion, while JNCL has a major commitment to the Foreign Language bill and the ADEA should they be re-introduced, it may also be wise to look at ways to improve legislation already in place. New legislation that will have the greatest chance of becoming public law in the 99th Congress will be legislation without price tags, such as Congressman Panetta's Bureau of Language Services bill combining and upgrading the translation and interpretation services already in existence, or Senator Simon's amendment to the U.S. Code providing extra points on the Civil Service exam for individuals with language skills. I anticipate that the major improvements in the next two years will come in the states. If we can protect the programs created or upgraded by the 98th Congress, preserve language and exchange funding at current levels, get the Higher Education Act reauthorized with a strong Title VI, and keep the current favorable public mood toward language study alive, we will have done very well in the 99th Congress.

IV

MACHINE TRANSLATION:

A LONG-TERM SOLUTION

MACHINE TRANSLATION: DEVELOPMENTS AND PROSPECTS

Robert A. Russell

Encouraged by advances both in computer technology and in linguistic theory, a number of scholars in the late 1950s and early 1960s began (in retrospect, naively) to talk seriously of the possibility of developing computer software capable of delivering "fully automatic high-quality translation" (FAHQT) from one natural language to another. With the enthusiastic support of the U.S. Government, a number of machine translation (MT) research projects were begun toward that end.

Having perhaps overestimated the power of the computer and, more importantly, having underestimated the enormous difficulty of explicitly characterizing the grammar of natural language, these early attempts to achieve FAHQT, while contributing greatly to our knowledge and experience, failed to produce the idealistic results anticipated by those funding the projects. The death knell for a number of promising MT projects came, of course, in the form of the largely negative report of the Automatic Language Processing Advisory Committee (1966) -- the infamous ALPAC report. The heady optimism and unrealistic promises born of an intoxication with technology had given way, in this country at least, to disillusionment and a critical reassessment of the potential role of computers in translation.

Following a period of relative inactivity, however, the past several years have seen a resurgence of interest in MT, fueled in part by the availability of faster, more powerful, less expensive hardware and by the burgeoning demand for language translation to facilitate the transfer of vast amounts of technological information. Meanwhile, our expectations of MT have become more realistic: few serious researchers still believe that FAHQT is feasible, now or even in the near future. Most have resigned themselves to the inevitable requirement of human intervention at some point in the MT process -- to the notion, in other words, of machine-*assisted* translation.

With the growing perception in the 1970s that MT was on the verge of becoming commercially feasible, the locus of activity shifted increasingly from the laboratories of academia to the offices of small corporations (SYSTRAN, Logos, Weidner, ALPS, etc.), while the emphasis in research and development began to shift from the former (research) to the latter (development). Commercial developers of machine-assisted translation systems understandably concerned themselves less with theoretical questions of parsing and semantic analysis than with bottom-line considerations of cost efficiency of throughput, volume break-even points, pay-back periods, and the like.

At roughly the same time that MT research was commencing in the United States and Europe, similar research was beginning to take place in Japan as well, at institutions such as the Electrotechnical Laboratory, Kyoto University, and Kyushu University (Nagao 1982:230 ff.). The Japanese interest in natural language processing (NLP) has continued to grow steadily since that time until the present, to the point where there is probably more activity in MT research and development in Japan now than in any other country (Nagao, interview in Shuukan Kompyuutaa Waarudo, 10/29/84). NLP/MT research has become even more important to Japan in the last few years, with its incorporation as a basic research and development theme within the Fifth Generation Project (cf. Brooking 1984, Institute of New Generation Computer Technology 1983).

Developments in Machine Translation

Let us consider the development of MT systems in general over the past several years, turning our attention in the process to more specific issues of MT development involving Japanese and English.

All approaches to MT are, of course, ultimately concerned with source languages and target languages and with the development of algorithms for transforming structures from the former into the latter. Of course, however, there is considerable diversity with respect to the specific algorithms that have been, or may be, developed. In general terms, MT design strategies have been characterized as either "direct" or "indirect" (Hutchins 1978:122 ff., 1982:22 ff.).

As the name implies, earlier direct MT systems were, from the outset, designed for particular pairs of languages and generally incorporated only as much analysis of either language as was absolutely necessary to transform structures from the given source language (SL) into those of the given target language (TL). Direct systems were typically ad hoc, highly empirical, generally atheoretical, relatively lacking in systematic semantic (or even syntactic) analysis, and had to be virtually rewritten for each new language pair or direction, as suggested below:

Source Language		Target Language	
Russian	—— i ——>	English	
English	—— j ——>	Russian	
French	—— k ——>	English	
English	—— l ——>	French	etc.

More recent indirect systems, on the other hand, are characterized by increasingly systematic syntactic and semantic analysis of both SL and TL, by the relative independence of SL analysis from TL synthesis, and by the incorporation (in theory at least) of an abstract, intermediate level of representation -- variously referred to as a "metalanguage," an "interlingua," or "pivot" -- to which any SL may theoretically be reduced, and from which any TL may subsequently be synthesized, as suggested below (cf., e.g., Hutchins 1978, 1982, Saino 1983):

The metalanguage model presupposes the existence of an adequate theory of universal grammar, which, it may be argued, still does not exist. Thus, although general concepts of universal grammar have a long history, even the most sophisticated MT system designs have been largely unsuccessful in implementing such ideas (Hutchins 1978:130).

In practice, most indirect systems have taken a more modest, three-stage approach consisting of SL analysis, SL-TL transfer, and TL synthesis. In this approach, for any given language the SL analysis is substantially the same regardless of the particular TL, and the TL synthesis or generation for each language remains the same regardless of the particular SL, and only the SL-TL transfer component differs significantly according to the particular language pair, as depicted in the following illustration:

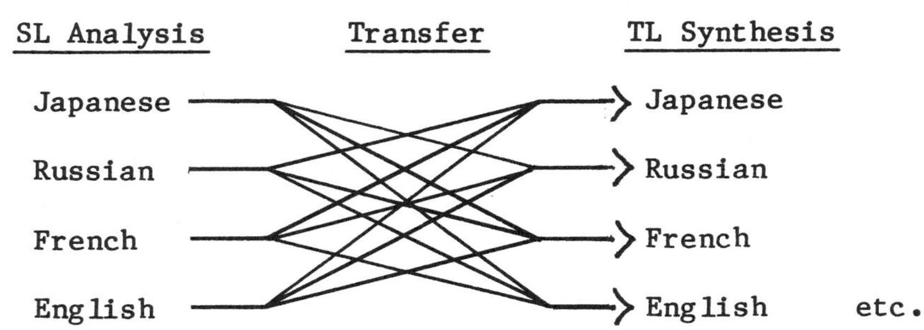

Actually, of course, there is considerable variation among transfer systems according to the depth of the linguistic analysis, ranging on a continuum from the most superficial, direct systems, in which almost the entire process is devoted to transfer, to the most indirect "metalanguage" (or "pivot") design, in which (theoretically) there is no transfer as such, but only analysis to a universal, underlying representation, and then generation of the corresponding surface representation in the TL. Most recent MT systems obviously fall somewhere between these two extremes, with transfer occurring at a relatively shallow syntactic level or at a deeper level involving increasing degrees of semantic analysis, as shown roughly below (cf. Saino 1983, Tucker and Nirenburg 1984, Hutchins 1982, etc.):

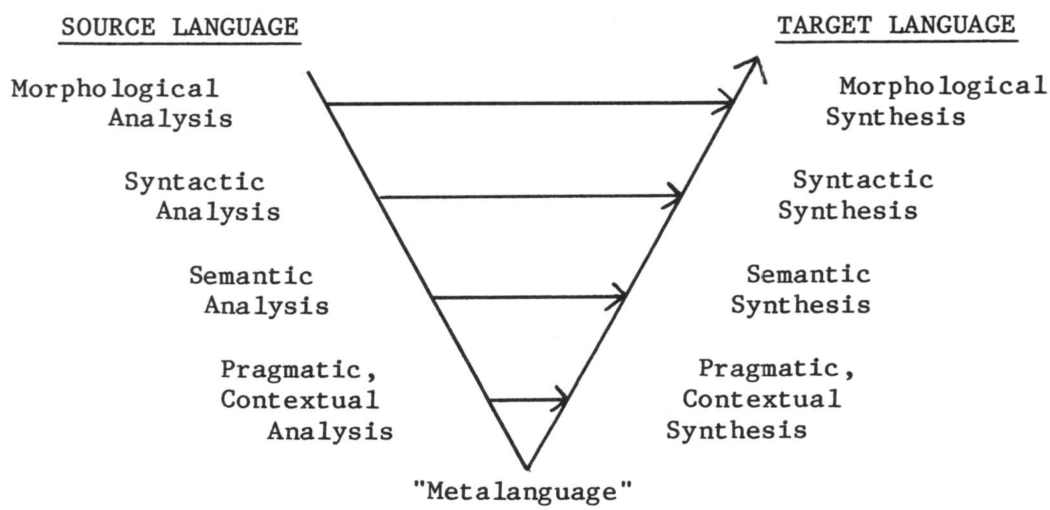

Most of the earlier MT systems involved translation only between pairs of Indo-European languages -- principally English, French, Spanish, German, and Russian. While automatic translation even between closely related languages is no trivial undertaking, the relative similarity of those languages to each other, both syntactically and semantically, permitted MT developers the luxury of taking a relatively superficial, direct approach, producing moderately comprehensible output with some syntactic analysis augmented by a minimal amount of lexical feature analysis.

In developing systems between languages as structurally dissimilar as English and Japanese, however, one is practically forced to pursue the linguistic analysis to a deeper level than might be required for more closely related languages. In the analysis of Japanese, for example, one must deal with problems such as the following:

1. Noun phrases that in English would normally be realized as pronouns are frequently deleted altogether in Japanese, recoverable only by semantically analyzing not only the sentence in which the ellipsis occurs, but usually also sentences in the preceding context as well.

2. The topic-comment structure and relatively free word order of noun phrases in Japanese (vs. the typical subject-predicate structure and relatively fixed order of noun phrases in English) makes a systematic semantic analysis all the more important.

3. The absence of morphological articles (a, an, the, etc.) in Japanese requires, in a Japanese-English system, that they be inferred from the context of the discourse.

4. Mapping between the lexical structures of any two languages is a complex matter, but the dissimilarity that exists between the lexical structure of Japanese and that of English is of an entirely different order of magnitude when compared, for instance, with that between most pairs of Indo-European languages.

5. The fact that normal written text in Japanese is largely unsegmented leads to a greater requirement for syntactic and semantic analysis just to arrive at a proper segmentation of the text into words, and so on.

In most of the commercial MT systems involving only Indo-European languages, SL-TL transfer occurs at a relatively superficial syntactic level. Most of the recent systems involving Japanese, however, pursue the analysis to a somewhat more abstract level. While Fujitsu's ATLAS I and Hitachi's ATHENE systems have been characterized in the literature, for example, as being relatively direct, involving transfer at a very superficial syntactic level, Toshiba's TAURAS system and the systems sponsored by Japan's Science and Technology Agency seem to involve relatively more complete syntactic analysis and transfer, with a fair degree of semantic analysis as well. The VENUS system of Japan Electric, the ATLAS II system developed by Fujitsu, and the case-grammar-oriented LUTE system developed by Musashino Electrical Communication Laboratory have been described as involving relatively full semantic analysis and transfer (Saino 1983). Little is publicly known concerning details of the method of analysis employed by the Bravice Japanese-English system, which has attracted a great deal of attention since its formal announcement last summer, nor is a great deal known concerning the analysis employed in SYSTRAN's Japanese systems.

With regard to quality of output, it is, of course, dangerous to generalize, especially when the only basis for judging the quality consists in limited samples provided by the developers themselves. Based, however, on the limited evidence that is available, it is probably safe to say that there is not currently a significant difference in overall quality of output between systems involving Japanese and English and those involving only Indo-European languages. It seems that the generally higher level of sophistication of the analyses employed in the Japanese systems is approximately offset -- indeed, necessitated -- by the increased difficulty of dealing with languages as structurally dissimilar as Japanese and English.

Among systems dealing only with Japanese and English, it is likewise premature to make any objective statement about the relative quality of the systems that are now, or will soon be, available commercially. The pattern of recent system announcements, however, clearly indicates the mounting intensity of the competition for the potential MT market in Japanese and English, in which quality will surely be a key issue:

System Developer	Language Direction	Date Available	
Bravice	Japanese-English	June 1984	
Fujitsu (ATLAS I)	English-Japanese	July 1984	
SYSTRAN Japan	English-Japanese	December 1984	
SYSTRAN Japan	Japanese-English	April 1985	
Fujitsu (ATLAS II)	Japanese-English	June 1985	
Bravice	English-Japanese	---	
Hitachi	English-Japanese	---	
Hitachi	Japanese-English	---	
Musashino (LUTE)	Japanese-English	---	
Musashino (")	English-Japanese	---	
IBM Japan	English-Japanese	---	etc.

Although some developers, for marketing reasons, might prefer the ambiguity of the current situation, a comprehensive (and independent) performance evaluation of the available systems would, at some point, be of great service to potential end-users.

Machine Translation Prospects

The growing commercial potential of machine-assisted translation which, in part, motivated the resurgence of interest in the field during the 1970s, also contributed to a growing preoccupation with the idea that "faster" and "bigger" were "better" -- that if the raw translation could be churned out fast enough, the sheer volume and speed of the output would, in terms of throughput efficiency at least, compensate for its frequently low level of quality or comprehensibility. Unfortunately, the general quality of the output has, in many applications, been well below the utility threshold for all but the most forgiving post-editors (Melby 1982:216 ff.).

Similarly, some have been seduced by the notion that the key to higher quality in MT output lay principally in larger memory capacities, more numerous grammatical rules, and larger dictionaries. To be sure, "bigger" has indeed led to some degree of improvement in overall quality. It is also encouraging to see that, with advances in hardware, the level of quality that a few years ago could only have been obtained on expensive mainframe and minicomputer hardware is becoming increasingly feasible on microcomputers.

The real key to further significant increases in MT quality and efficiency lies, however, not so much in larger stores of memory or in faster, more efficient computation, but in more insightful representation. By representation I refer to the specific models of syntactic, semantic, and pragmatic description employed in the processing of natural language, as well as to the general theoretical understanding of natural language that underlies those models. As Kay notes (1980:6), "truly significant gains in efficiency invariably come from adjustments to the algorithm itself, that is, to the overall strategy that the program employs."

As is true of many other fields, initial successes in the automatic processing of natural language come with almost deceptive ease, while subsequent improvements seem to come only at increasingly greater cost. This condition may be expressed graphically by viewing gains in MT quality as a function of resources expended:

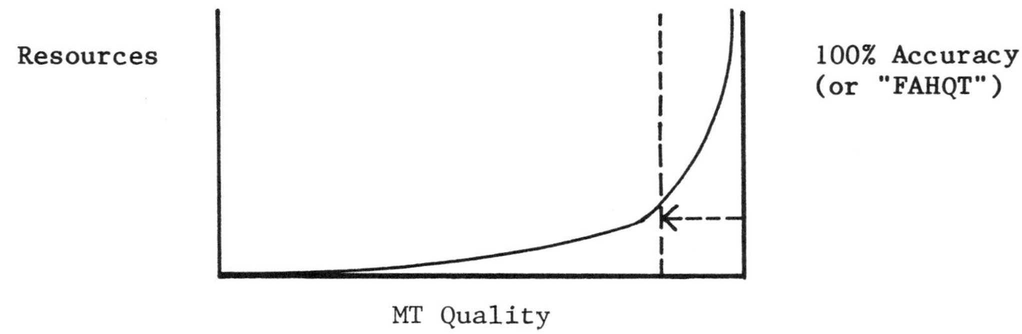

Assuming the existence of a reliable accuracy metric against which to measure gains in MT quality, experience suggests that the time, money, and manpower required to arrive at a level of approximately 80 percent accuracy, while considerable, are very small indeed when compared with the resources that will yet be required to approach the level of 100 percent accuracy. And it is precisely in the region where the resources/gains slope rises most steeply that increasingly sophisticated representation becomes critical. There is thus a point of diminishing returns beyond which further significant gains in MT quality are likely to come only as the result of an organized, long-term commitment of resources to basic research.

Of course, the asymptote hypothetically representing 100 percent accuracy in the above graph may, in effect, be shifted to the left by restricting the domain of natural language to be processed. Obviously, the further one restricts the domain of natural language, the more feasible 100 percent accuracy becomes in relation to current resources and technology.

A relatively successful example of a system developed for a highly restricted domain is the METEO system, an English-French MT system that has been integrated into the Canadian network for the transmission of weather forecasts (Thouin 1982:39 ff.). The system was developed by the TAUM group (the University of Montreal's Automatic Translation Research Team) with the support of the Canadian government, and it has been in regular use since May, 1977, yielding accurate translations, with no post-editing, approximately 80 percent of the time. METEO's success must, of course, be attributed in part to the fact that the system deals with a highly restricted subject area, much of the terminology and structure of which is relatively predictable.

An even more extreme case of restricting an MT system's domain of application is Kyoto University's system for translating scientific and technical titles from English into Japanese, which also seems to have achieved a high level of accuracy (an error rate of less than 5 percent, according to Nagao 1982:235) with only a very superficial level of linguistic analysis and transfer.

Most commercial MT developers are in fact very careful to note that their systems are intended only for the translation of technical texts, instructional manuals and the like. And while there may indeed be significant differences between a given sublanguage and its superordinate natural language, there is, in general, a strong tendency to overestimate the degree to which scientific and technical texts differ from, and are more tractable than more general texts -- both syntactically and semantically.

The quality line may also be shifted leftward by pre-editing or otherwise restricting the text at the point of input. Using the SYSTRAN system, Xerox claims "a five-to-one gain in translation time for most texts" (Ruffino 1982:57 ff.). This gain comes at the cost, however, of requiring a high degree of control over the input text, a cost that is unacceptable in many applications. Xerox writers using MT are required to produce texts in what they refer to as Multinational Customized English (MCE), which "involves a limited vocabulary,...[and] a set of writing rules which encourage a clear, concise English and a minimization of ambiguities" (Ruffino 1982:59).

What does the future hold for MT research and development? On the one hand, translation demand in many areas continues to exceed the supply of qualified translators by an ever widening margin, and translating costs continue to climb. If one could automatically process even only a small percentage of the manuals, catalogs, and technical documents currently being translated manually in Japan, the cost savings would amount to millions or even hundreds of millions of dollars per year (cf., e.g., the Japan Electronic Industry Development Association's survey, 1982:258 ff., and Saino 1983:55).

On the other hand, however, MT vendors continue to encounter skepticism and resistance to their product, despite the fact that current systems in most cases really are substantially improved over earlier ones. Superficial bottom-line considerations notwithstanding, potential users of MT continue to be most sensitive to quality. For many, the 80-85 percent accuracy promised by the vendor is not sufficient; for some, anything less than 95 percent accuracy is not even worth considering (von Alten 1984:35).

In light of the above discussion, some directions for future MT research and development might be suggested. In the short term, at least two different strategies seem promising: the development of systems that are highly application-specific, and the refinement of computer aids for translators. The recommended strategy for the long term, of course, would be to increase organizational and financial support for basic research in artificial intelligence and natural language processing in general, and in machine translation in particular. Let us, in conclusion, discuss each of these points briefly.

Most MT developers seem convinced of the importance of tailoring their systems to specific applications, as evidenced by their development of large, specialized dictionaries with, in some cases, the capability of "threading" the various system dictionaries at execution time, according to the subject area of the particular text. As discussed above, probably the least costly way to approach FAHQT is to restrict the domain of the language to be processed. Up to a point, the prospects for an MT system's success can be expected to vary directly with the degree of specificity of the domain to which it is applied -- whether it be weather reports, technical paper abstracts, patents, medical reports, contracts, or whatever. An MT system is, in a sense, a very complex expert system, and the more specialized it is, the more likely it is to be effective. Too many developers have spent too much of their limited resources trying to develop generalized systems for too many language pairs, instead of trying to identify more specific needs and focusing their resources on fulfilling those needs.

The need for computer aids for translators has long been recognized (cf., e.g., Kay 1980, Melby 1982). "Machine-assisted human translation" (MAHT) -- as opposed to "human-assisted machine translation" (HAMT) or "fully automatic machine translation" (FAMT) (Kay 1982:74 ff.) -- refers to the translator's use of sophisticated multi-lingual word processing with integrated, on-line dictionaries in a true workstation environment. While most MT systems obviously provide word processing and access to dictionaries, their design typically centers on the MT program under certain user-definable quality conditions. MT developers frequently cite the higher throughput efficiency of post-edited MT relative to <u>manual</u> translation. A more relevant comparison of efficiency might be that between post-edited MT and a translator's output using a well-designed system of machine aids.

Finally, we have alluded to the need for continuing basic research if significant advances in MT quality are to occur. An even more basic question remains, however: Is MT worth the long-term investment? Japanese government and industry evidently think so, judging again from the prominence accorded MT research in the Fifth Generation Project. Can the primary objective of the project's "intelligent interface function" -- namely, "to provide computers with a linguistic ability close to that of man" (Institute of New Generation Computer Technology 1983: 396) -- really be achieved within the time-frame of the project? If the project fails to achieve its ambitious goals for MT, is there a Japanese "Arupakku" on the near horizon? The answers to these and other questions will soon be known, but even if (as seems likely) the project does not achieve all of its MT objectives, it will still have made significant advances in the process.

What is the involvement of the United States in basic research of this type? With regard to the Fifth Generation Project in general, it has been stated that the "most important thing to say about the American response is that there isn't one; at least not a nationally coordinated project" (Brooking 1984:31). While there may not be a project of national scope, America has, nevertheless, been heavily engaged in all areas of basic research relating to natural language processing -- at universities (such as MIT, Stanford, and elsewhere) as well as in major corporations (IBM, Xerox, Bell Labs, and elsewhere) and consortia, both academic and industrial. The federal government's support of basic research has also continued to grow: in spite of budget cuts elsewhere, spending for R&D under the current administration has grown by 50 percent (to approximately $53 billion this year), and the budget for basic research ($8 billion dollars this year) has averaged double-digit increases each year (_Business Week_, 1/21/85, p. 108E), much of it in areas of at least indirect relevance for MT.

Conclusion

Cautious optimism concerning the future of MT seems justifiable. We have seen significant advances in both basic and applied MT research over the past several years. Much remains to be done, however, before MT of _consistently_ high quality can be expected for any but the most specialized texts. Further significant increases in the quality of MT output are more likely to occur as the result of continued basic research than as the result of improved engineering. Until significant improvements in fully automatic MT do occur, the greatest hope for broadly based gains in translation efficiency would seem to lie in the refinement of machine aids for translators and in the tailoring of MT systems to specific applications to the extent possible.

With respect to MT involving Japanese and English in particular, the focal point of activity is clearly in Japan, both in terms of hardware as well as software. There remain, however, a number of areas of natural language processing research in which there is room for both friendly competition as well as significant cooperation between the United States and Japan.

REFERENCES

ALPAC. 1966. Language and machines; computers in translation and linguistics. Washington, D.C.: Publication 1416, National Academy of Sciences.

Brooking, A. 1984. The Fifth Generation game. Expert systems: Principles and case studies, ed. by R. Forsyth, 18-35. London: Chapman and Hall.

Hutchins, W.J. 1978. Machine translation and machine-aided translation. Journal of Documentation 34.119-59.

------. 1982. The evolution of machine translation systems. Practical experience of machine translation, ed. by V. Lawson, 21-37. Amsterdam: North-Holland Publishing Co.

Institute of New Generation Computer Technology. 1983. Outline of research and development plans for Fifth-Generation Computer Systems. Reprinted in Byte 8.11.396-401.

Japan Electronic Industry Development Association. 1982. Kikai honyaku shisutemu no choosa kenkyuu. Tokyo.

Kay, M. 1980. The proper place of men and machines in language translation. Palo Alto: Xerox Palo Alto Research Center Report.

------. 1982. Machine translation. American Journal of Computational Linguistics 8.2.74-8.

Melby, A. 1982. Multi-level translation aids in a distributed system. COLING 82, ed. by Jan Horecky, 215-20. Amsterdam: North-Holland Publishing Co.

Nagao, M. 1982. A survey of natural language processing and machine translation in Japan. Kikai honyaku shisutemu no choosa kenkyuu, 230-40. Tokyo: Japan Electronic Industry Development Association.

Ruffino, J.R. 1982. Coping with machine translation. Practical experience of machine translation, ed. by V. Lawson, 57-60. Amsterdam: North-Holland Publishing Col.

Saino, T. 1983. Jitsuyooka e ririku suru 'shizen gengo shori gijutsu'. Nikkei Computer 3.21.55-75.

Thouin, B. 1982. The METEO system. Practical experience of machine translation, ed. by V. Lawson, 39-44. Amsterdam: North-Holland Publishing Co.

Tucker, A.B., Jr. and S. Nirenburg. 1984. Machine translation: A contemporary view. Annual Review of Information Science and Technology, ed. by M.E. Williams, vol. 19, pp. 129-60. White Plains, NY: Knowledge Industry Publications Inc.

von Alten, J.W. 1984. Translators gain fluency. Infoworld 8.13.35-7.

FUJITSU'S MACHINE TRANSLATION SYSTEM

Tatsuya Hayashi

Overseas sales account for more than 20 percent of all the sales of Fujitsu Laboratories Ltd. Accordingly, we have to translate manuals and brochures into foreign languages for overseas shipping, and the total cost of this material can reach as much as two billion yen per year. The increasing costs of translation in the field of international marketing has been a growing problem for most of the large companies in Japan. In an effort to reduce this translation cost, we started MT research seven years ago. We also wished to provide a product that would attract customers who were current or potential users of FACOM.

Our ultimate goal was to provide easier and quicker access to more scientific and technical information in Japan by overcoming language barriers through machine translation.

Historically, while the Japanese eagerly gathered information from overseas, the reverse was not the case. We have now become aware of the necessity for mutual information exchange, and for this, an MT system is considered an excellent tool.

Almost all the computer makers in Japan are developing Japanese-English machine translation systems. So far, only two of them are on the market: Bravice and Fujitsu.

Fujitsu announced their system to the press on September 20, 1984, and on the same day started shipping an MT system called ATLAS/I, which is an English-Japanese translation system. A Japanese-English version called ATLAS/II will be available in June 1985.

Other companies announced their MT systems to the press during the fall of 1984, but the actual availability dates were not given. The Science and Technology Agency of Japan sponsored a national project on MT research within the period from April 1982 to March 1986. (See Fig. 1 in Appendix.)

Fujitsu's ATLAS machine translation system won the "best product" prize of 1984 from Nikkei, one of the largest newspapers in Japan. These prizes were given to only 15 products from among 20,000 entries.

Before producing the commercial MT system, Fujitsu had developed an MT system as one of the customized application programs for a FACOM user. This is the RIPS Computer Center, which is shared among several national research institutes. In response to the center's request that Fujitsu develop a title translation system, we developed an English-Japanese machine translation system based on the prototype system developed by researchers at Kyoto University. This system is connected with Fujitsu's information retrieval system, FAIRS, and has been used for title translation of the papers retrieved by FAIRS since Spring 1981. (See Fig. 2 in Appendix.)

In September last year, we announced two new systems: ATLAS/I, English to Japanese translation system, and ATLAS/II, Japanese to English translation system. These systems can be utilized for translation of technical and business documents. In addition to these products, we are promoting several joint projects with overseas organizations for multi-lingual translation, which will be described later in this paper.

Approaches to Machine Translation Systems

There is no standard classification of machine translation systems. Each developer of MT systems has a different approach. Fujitsu is now proposing the standard classification shown in Fig. 3 (see Appendix), and hopes that this classification will prevail. For analyzing the source text, more and more complex information is used. The vertical line shows the relation between the amount of the information used and the understandability of the text by the system. The quality of understandability depends on the amount of information used.

Along the horizontal line, we find the words "direct" and "transfer." "Transfer" means converting source text into an intermediate representation and then generating target text from the representation. "Direct" means generating target text from source text without employing an intermediate representation. According to this classification, the RIPS and ATLAS/I systems developed by Fujitsu should be called syntax direct, and ATLAS/II should be semantic transfer.

At the bottom of this triangle, our finite goal of this approach, "pivot," is shown. This approach is to translate text through the universal representation, which is not directly connected with any specific natural language such as Japanese, English, and Arabic. Some linguists are pursuing this approach based on, for example, the Montague theory.

It is not clear whether or not the Fifth Generation project will enter into the MT research and development. Currently, MT research is said to be less attractive to Fifth Generation researchers because private companies have already developed MT systems. However, there is still enough room for the Fifth Generation project to devote efforts to more advanced approaches to MT.

The ATLAS System

The ATLAS system is supplied to users in a software package working on general-purpose computers, not stand-alone machines.

ATLAS/I needs a large computer: M-series or a super mini-computer, S-series. M-series is an IBM compatible machine, and S-series is a mini-computer incorporating the same architecture as the M-series. The date of shipping ATLAS/I for a large M-series machine was July 1984. The date for smaller M-series will be March 1985, and that for S-series, June 1985. ATLAS/II is also scheduled to be shipped in June 1985. (See Fig. 4 in Appendix.)

The rental price of ATLAS/I is 350,000 yen per month; that of ATLAS/II is 550,000 yen per month. The basic system of ATLAS includes a 50,000-word dictionary. As an option, we provide a technical term dictionary with 250,000 entries for 150,000 yen per month.

ATLAS/II is a translation system of Japanese into English, based on a semantic transfer approach. The Japanese language has unique characteristics: the word order is relatively free, except that the predicate comes last; the subject of a sentence is often omitted; the concepts of singular and plural nouns are not clear; and articles and prepositions are not used. (See Fig. 5 in Appendix.) These characteristics make the Japanese analysis complex. The analysis should employ not only syntactic but also semantic information. ATLAS/II has the following dictionaries: a semantic dictionary that stores meaning relations; a structure conversion dictionary that converts a sentence structure specific to Japanese into a structure suitable for generating English; a co-occurrence relation dictionary that selects a suitable English word in generation. For instance, Japanese can use a word "naosu" in many instances, but in English, a suitable word must be selected according to the verb's object; that is, "update" for the object "program," or "repair" for "device," or "correct" for "error," and so forth.

The Cost Efficiency of ATLAS

Generally speaking, conventional translation consists of the manual translation and the typing of the manuscript. Automatic translation consists of three steps: pre-editing of the input text, MT translation, and then post-editing of the translated text. (See Fig. 6 in Appendix.) We expect that the MT system will allow a 60 percent reduction both in time and in cost. The conventional procedure requires 60 minutes for translating one page, 45 minutes for manual translation, and 15 minutes for formatting. The MT system, however, needs less than 25 minutes: 10

minutes each for input and post-editing, and at most one minute for MT system on M380. (See Fig. 7 in Appendix.)

The total cost of conventional translation is about 5,000 yen, including 3,500 for manual translation and 1,500 for typing. The MT system costs about 2,000 yen, including 1,500 for pre- and post-editing and 500 for MT translation.

If, therefore, you have translations of more than 1,000 pages per month, it would be advantageous for you to use the MT system with a M340R computer; if you have even 200 more pages of translation, it would be cost-efficient to acquire another terminal.

MT Development for Multi-Lingual Translation

Fujitsu is working on a machine translation system for language pairs other than English-Japanese, aiming at multi-lingual translation. One of these is a Japanese-Korean translation system. A long-term research committee of economic cooperation to organize the joint projects in various fields has been organized between Korea and Japan; the 3-year project began in March 1983. This research is organized by Fujitsu Japan and KAIST of Korea. The detailed research is done in FKL, a subsidiary of Fujitsu in Korea. The target text of machine translation is technical papers. (See Fig. 8 in Appendix.)

Another project is a Japanese-to-German MT system. A technological cooperation committee has been organized by the Japanese and German governments, with the Science and Technology Agency of Japan and BMFT of Germany participating. The partner of Fujitsu is Stuttgart University, and research on the 2-year project began in July 1983. We expect that the period of research will be extended. (See Fig. 9 in Appendix.)

We are also considering similar projects with other languages, especially those spoken by many people, such as Spanish and Chinese.

Future Prospects of MT Systems

The syntax-based technology has already been put into practice. The next step will be semantic-based technology. Syntax-based systems will reach a plateau about 1990, and after that will be replaced by semantic-based systems. The contextual knowledge-based system, which has almost the same capability as an expert human translator, is now nearing the research and development stage, but these technologies will not be in use until the 1990s. (See Fig. 10 in Appendix.)

Fujitsu has also been developing various important technologies in the field of basic software. In the 1960s we developed assembler and compiler technologies; in the 1970s, TSS and batch processing technology; and in the 1980s, database and computer network architecture.

Our efforts thus far, as well as those of other companies, have made Japan second only to the United States in the computer market. (See Fig. 11 in Appendix.)

However, natural language is not understandable by computers. Current man-machine interface is still at a machine level. If we could make computers understand our languages and make humanlike man-machine interface, then computers would be more friendly to humans and more useful in daily life. These facts lead us to believe that natural language technologies such as MT, as well as knowledge engineering in AI, are fundamentally important projects.

Cooperation with Public Organizations

IPA is a special organization under MITI, and researchers from Fujitsu are included in the group. IPA helps linguists compile Japanese dictionaries.

MITI also established ICOT to pursue the Fifth Generation project. In the ICOT project, Fujitsu and other computer manufacturers are engaged in compiling Japanese-Japanese, Japanese-English, English-English, and English-Japanese dictionaries.

Another government institute, the Scientific and Technological Agency, launched an MT project with JICST to develop a technical term dictionary.

We anticipate that the results of these projects will be available for our MT systems. (See Fig. 12 in Appendix.)

Cooperation with Overseas Organizations

We have developed a Japanese-English machine translation system by using our own technologies. However, in order to enhance the quality of translation, participation of native speakers is essential. In this regard, we are seeking contacts with overseas organizations. I believe that machine translation is one of the most appealing projects in the field of overseas cooperation.

We can see some constructive and positive signs for overseas cooperation. The status of the Japanese language in world communication is being enhanced thanks to the advancement of scientific technology in Japan. Today's conference is an evidence of this phenomenon. As development of a worldwide network of computers becomes a reality, cooperation in developing machine translation systems becomes easier. Also, progress is being made for computers to deal with the languages of the developing countries. At Fujitsu, for instance, it is possible to input Korean and Chinese characters into computers. Finally, the Minister of

Posts and Telecommunications, at a meeting in Cannes, France, last summer, proposed research on vocal translation through telephone systems. (See Fig. 13 in Appendix.)

International Japanese

The Japanese language has to be awakened from a long sleep and has to be opened to the world. In order to realize this, we have to convert "local Japanese" to "international Japanese." Although the Japanese speaking population is the sixth largest in the world, it is limited to Japan. The Japanese are a single ethnic group, and they share the same cultural background, the same life style, and the same sense of values. Communications in Japanese tend to be implicit, and the implicit information is augmented by background knowledge. A large part of the information is unuttered and hidden beneath the surface, like an iceberg in water. Explicitly spoken words are just a small part of the whole story. (See Fig. 14 in Appendix.)

This nature of the Japanese language makes it difficult to internationalize Japanese. What can we do now? The answer is simply to make Japan a multi-ethnic country. However, I am not talking about importing people from overseas. In Japan, we have many computers, but they are like dogs, horses and cattle because they do not understand natural languages. But as I have already discussed, making computers understand natural languages will be realized in the future, and then they will become another ethnic group in Japan.

When that happens, we have to describe things precisely, since we do not share the same culture and life style with computers. These efforts will make Japan and the Japanese people internationalized. Our urgent task is the development of these computers.

Figure 1

Present MT Developments in Japan

Company	System	Language	Date	Application	Approuch
Bravice	PAK-II	J → E E → J	Jun / 84 (on the market)	scientific and technical documents etc.	semantics
Systran Japan	SYSTRAN	E → J J → E	Dec / 84 (announcement)	scientific and technical documents	syntax
Hitachi	ATHENE	E → J	Sept / 84 (announcement)	economic news	syntax
	ATHENE/N	J → E	Oct / 84 (announcement)	scientific and technical documents	semantics
NEC	VENUS	J → E E → J	Nov / 84 (announcement)	scientific and technical documents	semantics
IBM Japan	–	E → J	Aug / 84 (announcement)	computer-output massages	syntax
Science and Technology Agency	Mu	J → E E → J	April / 82 | March / 86 (research)	scientific and technical documents	syntax
Fujitsu	ATLAS-I	E → J	July / 84 (on the market)	scientific and technical documents	syntax
	ATLAS-II	J → E	June / 85 (on the market)		semantics

Figure 2

FUJITSU's Machine Translation Systems

Approach	System	Source / Target	Status
Syntax Direct	RIPS	Title E → J	in Use
	ATLAS/I	E → J	on Sale (July / 84)
		J → K	Japan–Korea Joint Project
Semantic Transfer	ATLAS/II	J → E	to be released in Jun / 85
		J → G	Japan–Germany Joint Project

Figure 3

Approaches to Machine Translation

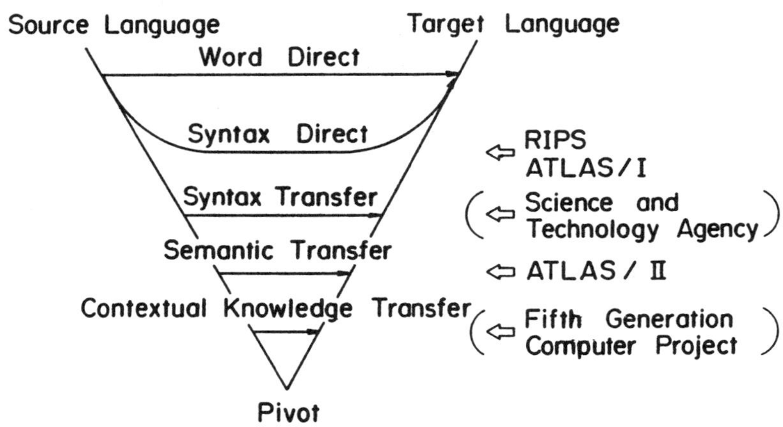

Figure 4

Outline of ATLAS system

		ATLAS-I	ATLAS-II
Language Pair		E → J	J → E
Environment	CPU (hardware)	M series, S series	M series
	OS (software)	MSP, FSP, OVIS/S	MSP
Price	Basic	350,000 yen/month	550,000 yen/month
	Optional	Technical term dictionary : 150,000 yen/month	
Shipping Date		MSP : July/84 FSP : March/85 OVIS/S : June/85	MSP : June/85

Figure 5

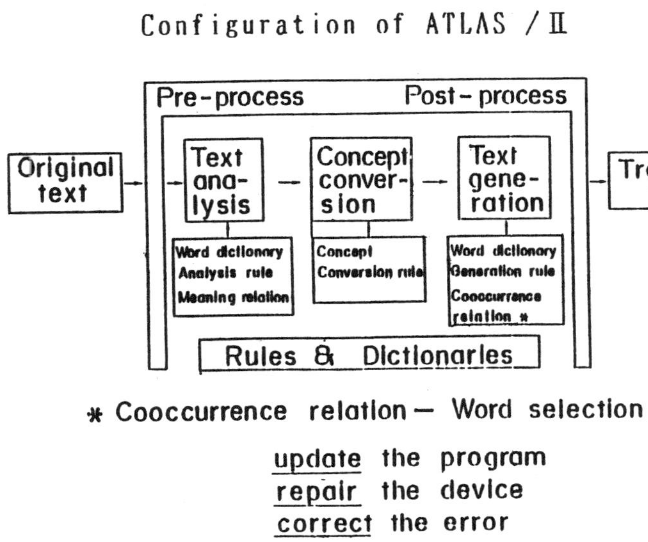

Figure 6

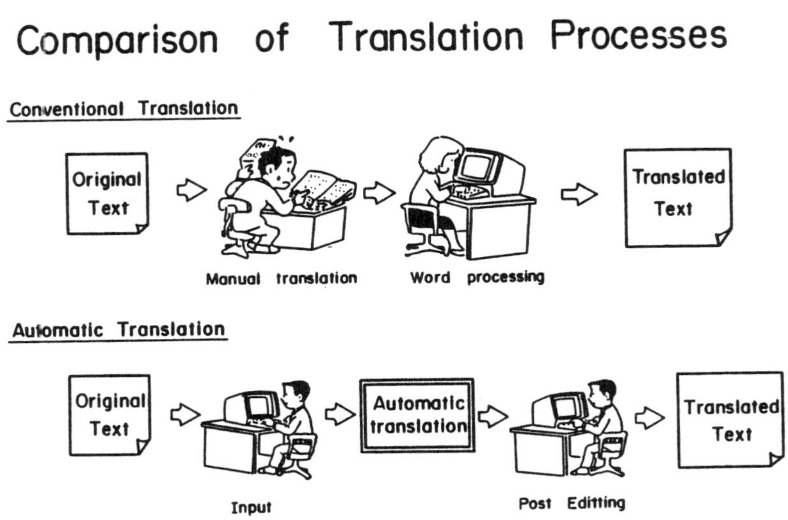

Figure 7

Performance of ATLAS

	ATLAS	Mannal Translation
Translation Time (A4, I page)	25 min. (input + translation + post editting)	60 min. (translation + formatting)
Cost (A4, I page)	2,000 yen	5,000 yen
Break-even Point ① M340R (basic) ② WDS+RP (additional)	I,000 pages/month 200 pages/month	—

Figure 8

Japan-Korea Joint Project
on Machine Translation

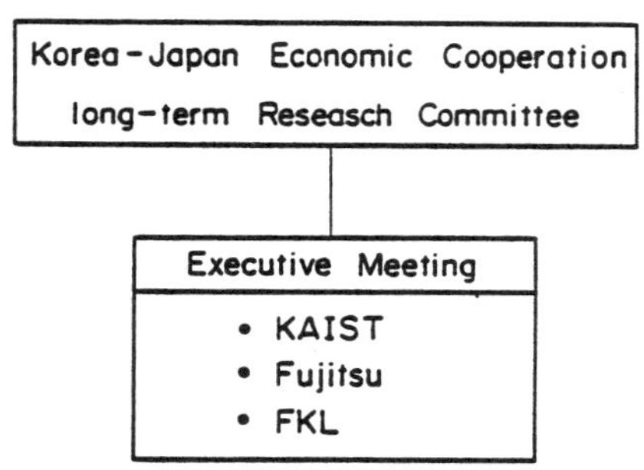

Reriod : Oct. 1983–Sept. 1986

Text : Scientifie and technical documents (JICST)

Figure 9

Japan-German Joint Project

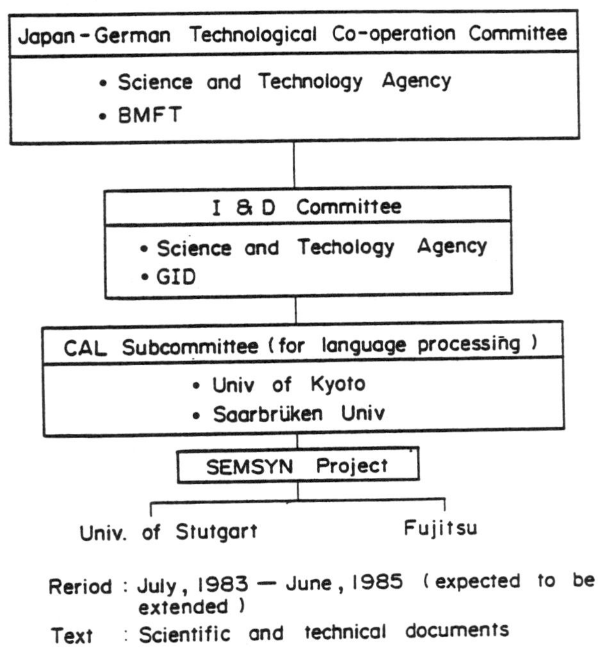

Figure 10

Translation Technologies and Their Commercializations

Approach	Information type G	Information type S	Information type C	Quality	Prospect for practical use
Syntax-direct / Syntax-transfer	○	–	–	○	at present; R & D → practical use; 80, 85, 90, 95
Semantic-transfer	○	○	–	◎	R & D → practical use; 80, 85, 90, 95
Context-transfer	○	○	○	◉	basic study → R & D → practical; 80, 85, 90, 95

G : Grammar
S : Semantic
C : Context

Figure 11

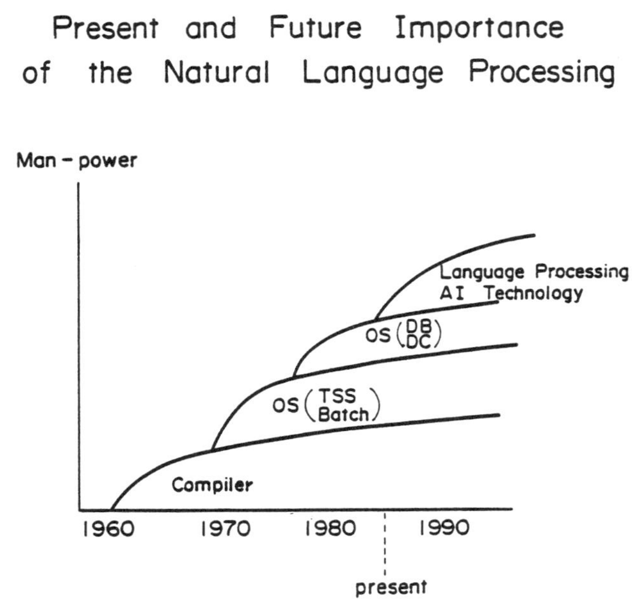

Figure 12

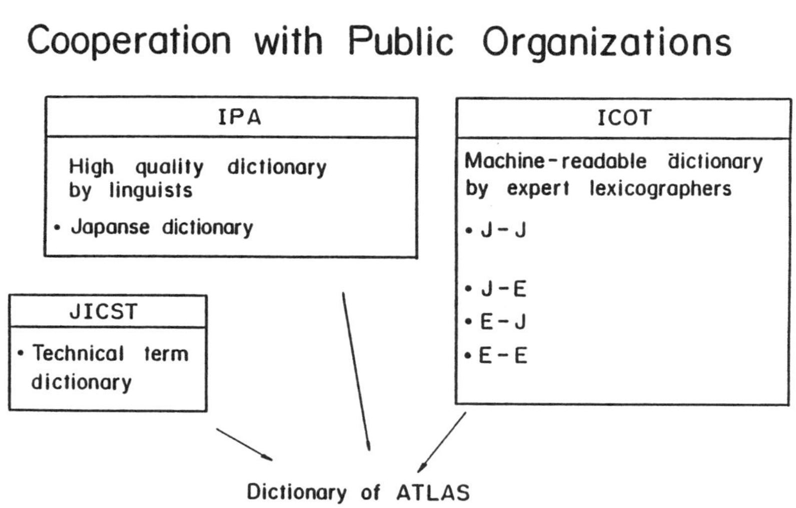

Figure 13

Overseas Cooperation

1. Higher international status of Japanese language
2. Creation of international network
3. Development of word processors for non-latin languages
4. Planning of vocal translation for telephone systems
   (Ministry of Posts and Telecommunications)

Figure 14

International Japanese

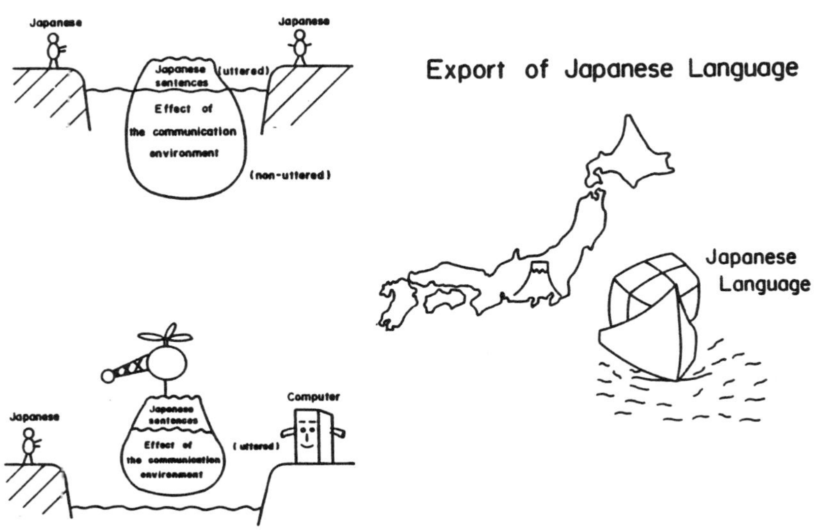

THE BRAVICE COMPUTER TRANSLATION SYSTEM

Takehiko Yamamoto

Although I am not a descendant of Admiral Yamamoto, who was shot down in the Pacific by the U.S. Air Force, some people, especially bankers, find a certain resemblance between me and Admiral Yamamoto, who used to love gambling. In their eyes, the attempt to develop a computer translation system between English and Japanese, which differs significantly from English in syntax and grammar, looked even more risky than gambling.

Today Bravice's computer translation system for translating from Japanese into English is here for your appraisal. Before entering into the details of the system, let me give you a short history of Bravice.

Thirteen years ago we started as a consulting company for Japanese industries who intended to establish themselves in Brazil. In the course of business development, we were obliged to undertake many translation jobs for our clients, and these translations became a significant source of revenue for Bravice. However, we soon encountered a shortage of capable translators. By nature, translation requires a high level of understanding and skill in a language, and a sound knowledge of the item in question. Such talent is difficult to find, especially at a specific time and in required numbers. It is my experience that any company that does a considerable amount of export business is struggling to obtain quality translation of its documents.

To improve the situation, we tried a computer-aided dictionary support system, which ended in total failure. Seven years ago, we started to approach computer translations by collecting information and basic data.

In 1979, we came across Weidner Communications, which at that time was located in Provo, Utah. Weidner had just completed a computer-assisted translation system between English and Spanish, and Bravice purchased a system, making it the second client of Weidner Communications (WCC). Two years later, we decided to let WCC develop under our auspices and advice a system for translation from Japanese into English, and the project began.

During the course of the development, I found it necessary to involve our staff more heavily in the project. The only way to achieve this seemed to be through acquisition of the majority position in WCC's equity. We made this acquisition in December 1982. Bravice's Japanese-English computer translation system was completed in May 1984, and the product was launched into the market the following month. Relative to this product, Bravice has since achieved sales, including software licensing, equivalent to approximately $4 million.

Philosophy of Development

When we announced our system in May of last year, we were the first in the world to introduce a machine translation system. The very next day, the Asahi Shimbun, one of Japan's leading newspapers, carried the story at the top of the first page, and it really ignited a spark. Reporters from various magazines and newspapers rushed to us, and today the magazines that had articles reporting Bravice and our new product pile up to a height of two feet!

After we gave the reporters a short demonstration, they requested trial translations of Japanese literature. A Japanese Hamlet showed up -- this shows how great their expectations were! However, our goal was to translate technical documents and journalistic reports, excluding areas like art or culture in general.

As most researchers in language processing assert, perfect machine translation cannot be expected at this stage of development in hardware and software. Based upon this understanding, the goal was set for development of a system that would produce a raw translation that could be edited without the translator's having to refer very much to the original source-language text. You probably agree that if such a standard was not met, then the economic feasibility of computer translation would be totally lost.

People tend to think that some day the computer will be a perfect substitute for a human translator. That probably will not happen, but this does not reduce the role and importance of computer translation.

Basic Features of the Software

Presently Bravice's Japanese into English software is written mostly in Digital Equipment Corporation's Fortran 77 and partly in Assembler language. The reason we have adopted Digital's Fortran is that it has a certain expanded function that facilitates the manipulation of strings.

Our codes stand basically for the so-called Transfer Method. The program begins with segmentation of the source text. Segmentation is a procedure in which Japanese words are separated from each other. Since the Japanese language does not have spaces between words, this procedure, which does not exist in machine translation for European languages, is an absolute necessity for us. In Bravice's system, the segmentation is conducted in the bottom-up method.

Morphological analysis, which follows segmentation, breaks down terms into ten parts of speech. Then syntactical and semantic analyses of the source text are conducted in an integrated method, which forms a characteristic of Bravice's algorithm. Another characteristic in the syntactical analysis is a step-down analysis, a maximum of six stages. This analysis eliminates the difficulty of finding a single correct analysis among possible answers given by parsers using the conventional augmented transition network. This means that our parser always gives us a single analysis of the source text.

Here I would like to mention the structure of the dictionary. The electronic dictionary built into the system consists of a core dictionary and satellite dictionaries that stand for specific fields. There is no limit on the number of terms that can be stored in the core dictionary or in the satellite dictionaries. At present, approximately 50,000 terms are in the core dictionary. Users have free access to the core and satellite dictionaries. In other words, users are permitted to update the dictionaries by themselves, except for some 8,000 basic terms in the core dictionary which are restricted. The combination of the core dictionary and satellite dictionaries (called multi-subject dictionaries) allows users simultaneous operation of dictionary look-up and dictionary updating. This is a very effective feature in the economy of dictionary maintenance.

Our electronic dictionary registers a term using twenty attributes. Each attribute can have 16 values at the maximum. Correct determination of attributes is one of the fundamentals for correct syntactical and semantic analysis.

Generation of the target text, following the syntactical and semantic analysis, occurs through the transfer stage, which handles discrepancies between structures of the source and target languages. Finally, inflection procedures complete the automatic translation.

Basic Features of Required Hardware

Since Bravice's software is written in Digital's Fortran 77, it runs on Digital's VAX series and PDP 11 series.

In addition to the VAX and PDP 11 series, we have the PAK 11 series, which emulates the PDP 11 series under the same operating system, that is to say, RSX-11 M-plus.

The above hardware requires Roman character into *kana-kanji* conversion software, which is provided by Digital or Bravice. This software facilitates entry of source text into the system. As you may know, kana is a phonetic character system specific to the Japanese language; kanji is an ideographic character system originally borrowed from the Chinese. The Japanese language is written using 50 kana and several thousand kanji symbols. Therefore, entry of source text is almost impossible without effective kana-kanji conversion software. I dare say that kanji characters used to be a greater barrier for U.S. computer companies in Japan than the notorious industrial policy. I am really convinced that many U.S. companies ignored, or at least underrated, the importance of the kanji problem, and most U.S. computer equipment still has less sophisticated Japanese language support than do Japanese computers. In addition to the Roman character to kana-kanji conversion software, a display and a printer that have Japanese support should be included in the hardware configuration.

The minimum requirements for Bravice's software are a 512 K core memory and 20 megabytes of external memory, including user areas. Depending upon the size of the dictionary required for a given user, the capacity of the external memory may need to be enhanced. Generally speaking, 5,000 terms necessitate one megabyte in the external memory.

So far, every other attempted computer translation system between Japanese and English has been tried on a mainframe computer. Bravice has introduced a more economical option for users without sacrificing quality of translation. The only sacrifice is translation speed, which actually is not a sacrifice for most users. In the case of VAX 750, the translations are generated at the rate of 2,500-3,300 words per hour for a single user, depending upon the nature of the source text.

Translation Procedure

Bravice's computer translation system follows the following procedure: First, a source text is input through a keyboard. If the source text is available on a magnetic medium such as tape or diskette, automatic input may be possible, resulting in savings in time and labor. A completely automatic vocabulary search follows, which identifies the terms not registered in the computer's dictionary. A user can skip the vocabulary search process and jump into the automatic translation process. However, skipping the vocabulary search is generally not recommended, since an unregistered term can confuse the syntactical and semantic analysis of the source text. Naturally, this would produce an incorrect or even meaningless translation.

If the vocabulary search shows a term that needs to be registered in the dictionary, the user proceeds to dictionary updating. Dictionary updating can be mastered usually in less than three weeks. In our experience, every user has a specific dictionary for his/her own use. This dictionary should be added to the system's electronic dictionary in advance.

After dictionary updating, automatic translation takes place. In this stage, there is no human intervention. When the translation is finished, the source text and the target text are shown, paragraph by paragraph, on the display.

Post-editing is the final touch to the raw translation generated by machine. Usually a native speaker does the post-editing, correcting grammatical errors and giving the raw translation more sophisticated sentence structuring if necessary. A post-editor who is a specialist in the field in question is desirable, even if a considerable portion of specific knowledge required for the translation is given by the system, including correct selection of terms. The post-edited translation may be printed out or transferred to magnetic tape or diskette.

Quality of Raw Translation

Quality of the raw translation is a key factor in computer translation, and appraisal of the quality involves many difficult problems. I have not yet seen a reliable method for evaluation of the quality of raw translation. Communication is an area in which 100 percent performance cannot be expected. A certain sentence might get a perfect mark from one person and an inferior mark from another person. Therefore, I would like to avoid attaching any marks to our system. What we say is that in most cases our raw translation transfers the meaning of the source text in a very understandable manner. We admit that certain types of source text will result in an imperfect transfer of the meaning. Such types of source text should go through pre-editing, in which the source text is rewritten into another type of sentence structure without changing the meaning. In general, pre-editing is performed for two reasons: to improve the quality of source text if there is confusion in syntax and logical development of thesis; and to overcome the difficulty in the system's handling of certain types of syntactical structure.

One difficulty in machine translation from Japanese into English is the proper handling of the relationships between modifying and modified words. The toughest problem is the relationship between nouns. This problem can be defined as one of correct determination of noun phrases, and it is an essential factor in technical literature. Bravice provides a solution to this difficult problem which no researcher has ever successfully solved in a simpler way.

The quality of the raw translation is very much influenced by the quality of the source text. This is especially true with Japanese into English. In other words, half the difficulties in translation from Japanese into English originate in the poor quality of the source texts. This is by no means an excuse, but rather it is an opinion shared by many people who have been seriously involved in Japanese to English translation.

In the Japanese educational system, a student has little opportunity to develop writing skills. This lack of opportunity has quite a bit to do with testing for upper level education. Some time ago, when I referred to our development of a Japanese-English translation system at a meeting of a computer society in Tokyo, almost all of the American translators did not believe such a system was possible. And two months ago, several translators from a certain translators' association in Tokyo visited us to test the translation of a source text on our system. The text was an extraction from a newspaper, a badly written extraction. The result was miserable. However, Bravice's system was not at fault. No native speaker of Japanese could reach a single interpretation of the text; everyone who read the text had a different understanding. In short, in discussing the quality of raw translation, one must appraise the source text before making a judgment.

Economy of Translation by Bravice's System

If the quality of the raw translation is acceptable, the next question is the economy of translation using Bravice's system.

It may be too early to give a cost reduction figure based on Bravice's own internal translations, since daily development and enhancement operations still compete with the commercial operation of the system. However, a rough estimation shows a 30 percent cost reduction, and further reduction, possibly up to 70 percent, would not be difficult given enhancement of the core dictionary and the specific field dictionaries.

The economy of machine translation is greater or less depending upon the application of the system. Translations may be classified into two categories: one includes outgoing documents such as technical reports, manuals, official papers, etc., and the other includes in-house papers that often need immediate translation. The former category always requires a complete finish in style and expression, but the latter just needs to be translated quickly, even with some sacrifice of the translation's quality. Such cases may be ideal for Bravice's system without post-editing. This certainly eliminates a considerable portion of the usual translation cost.

An invisible factor in cost reduction is the quality of consistency in translation. Large-volume translation usually involves many translators, inevitably resulting in inconsistent quality. There is no practical way for making the quality consistent except by machine translation. Moreover, in manual translations, the same word in the source text may be translated in different ways by the various translators. In some service manuals for a certain motorcycle, we found seven different sentences translating the same source sentence.

Future Enhancements

In less than three months, Bravice's software will run on any hardware supporting C language under UNIX. Thus the portability of the software will be improved to a great extent. The minimum requirements for the size of core memory and external memory will be unchanged.

In addition to greater portability, translation speed will also be improved significantly. When the software runs on a mainframe, the translation speed may exceed 100,000 words per machine per hour.

The size of the core dictionary will periodically be enhanced, and corresponding increases in the number of idioms will contribute to a better raw translation quality.

Beginning late in 1985, the maintenance of Bravice's Japanese into English system in the United States will be assumed by WCC in Chicago. We are very proud of the joint operation between Bravice and WCC both in development and in sales promotion. I am thoroughly convinced that cooperation across the ocean is indispensable for creation of a sound and effective computer translation system. We would not be able to demonstrate our system to you today without the full cooperation and support of WCC.

V

LANGUAGE TECHNOLOGIES: A MEDIUM-TERM SOLUTION

THE MIT ATHENA LANGUAGE LEARNING PROJECT: A VIEW FROM THE FUTURE

Claire J. Kramsch and Janet H. Murray

U.S. public opinion is becoming increasingly aware of the importance of foreign language study to meet the challenges of a multinational environment. Some are turning to the language teacher for quick fixes that will make possible economically desirable moves, and are even hoping that a new technology will eliminate the need for the human teacher altogether. This paper examines some innovative uses of the technology that will be available to us in the near future for language teaching. It also reconsiders the role of language as a tool for personal, social, and cultural communication, and the ways in which CAI can sharpen that tool.

At MIT, language teachers are privileged to have access to the world leaders in artificial intelligence, speech processing, educational computation, and interactive video. We are now able to take advantage of this community of unmatched technological expertise by virtue of Project Athena, a five-year, campus-wide educational experiment for the integration of computers into the curriculum. The computers, supplied by DEC and IBM, will be powerful, networked work stations, featuring bit-mapped graphics, 32 bit processing, enhanced memory and speed, running Berkeley Unix as the operating system. We also have available to us the DEC IVIS system for interactive video, and other IBM-based variants, as well as equipment for digital storage, retrieval, and analysis of speech. This system is, in our judgment, an accurate model of the kind of environment that will be available to foreign language teachers at prices comparable to today's widely used but much less capable micros.

In January 1985, the MIT Foreign Languages and Literatures Section embarked on a five-year grant from the Annenberg/CPB Project to use the resources of Project Athena to develop materials for teaching a four-course sequence in each of the five languages taught at MIT: French, Spanish, Russian, German, and English as a Second Language. These materials are to use artificial intelligence in natural language processing and to include interactive video and interactive audio components. They are to be adaptable through authoring systems for use by other teachers and for use in learning other languages. In fact the kind of artificial intelligence system we are using makes the substitution of languages relatively simple. We are limiting ourselves right now, however, to the five languages taught at MIT.

The materials that we are developing are meant to supplement a normal course sequence, to replace time spent with a textbook or with an audio tape, but not to replace classroom time. We are now engaged in designing a new Language Resource Center, including adjacent classroom space. These materials will be used in such an environment: a place where students can work singly or in small groups on exercises that are then followed up by classroom interaction. The language lab is seen as an extension of the classroom and a discovery-rich environment for the student to explore and interact with.

In addition to text-based exercises operating on artificial intelligence processing of typed-in sentences, we are developing a demonstration project with interactive video and a pronunciation/intonation-practice system using advanced digital speech equipment. Eventually, we expect these three kinds of media to be interconnected, so that we will be able to combine text-based exercises with interactive video and interactive audio.

In these first two years of our project, we are engaged in developing building block exercises that encapsulate representative engineering tasks and representative learning tasks. After two years, we will evaluate these building blocks in order to decide on how we will design a coherent series of exercises using these techniques.

Although we have only just begun the Annenberg/CPB grant period, we have received seed money from Project Athena for 18 months; we have therefore begun programming on two of these building blocks, and we have made much progress in planning and conceptualizing others. The basic design of our system rests on the Marcus parser, an artificial intelligence program that can parse a sentence based upon any set of grammar rules. Connected to this parser is a knowledge representation system that will allow us to "understand" as well as syntactically represent students' words. Included in this system is a series of error routines for recognizing spelling mistakes, typos, incorrect grammar, and culturally incorrect or semantically incoherent statements. The use of artificial intelligence allows us to do this kind of correction generically rather than require a teacher to anticipate every individual error a student might make.

Discourse Training Prototypes

We are currently developing in parallel three building blocks of a discourse training program corresponding to the three levels of languages taught at MIT.

The two programs currently in code are LINGO and ToPicks. LINGO (Language Instruction through Graphics Operations) is a complicated AI system, really the basis of much of our future work. It was designed as an undergraduate thesis by Edwin Seidewitz, under the supervision of Professor Robert Berwick of MIT's AI Lab, who is in charge of the technical side of our project. As an exercise, it works like this: A student is shown a picture of a dorm room with some representatively German objects in it, and is then introduced to a "poltergeist" with whom the student is invited to converse. The "poltergeist" will mess up the room following the student's orders, but in order to talk with the spirit, the student must follow certain rules of conversation. Our parser and knowledge representation are powerful enough so that we can accept sentences of much greater complexity than the "Kill troll with sword" variety of remark common in commercial adventure games. The messed-up room picture will be stored to be retrieved by another student who will be given the task of cleaning up the room.

This "poltergeist" uses natural discourse that includes such conversational strategies as requesting clarification if he does not understand a certain directive, suggesting grammatical or spelling repairs, giving listeners feedback, etc. The student thus learns to take the computer seriously and enters into negotiation whenever difficulty arises in the communication.

The second exercise currently in production is ToPicks. This is a rudimentary conversation in which a student trades one-word utterances with the computer, augmented by some rote sentence-length exchanges. Topics is currently built in an experimental version and is being tested with students. The ToPicks exercise works simultaneously as a review of German I vocabulary and as an exercise in topic-nominating and topic-building. Eventually we expect to expand it so that the players are trading phrases and sentences rather than just words.

A third exercise, Tacit-Turn, will be designed for the more advanced levels. The student has to pry information out of a reluctant informant, using tactful indirect questioning techniques. Direct WH-questions will only elicit scanty responses from the computer. By contrast, more empathetic, indirect statements, such as reactions (oh, really?), commentaries (that's too bad!), repetitions and paraphrases (you were sick?), interpretations and inferences (so you mean you had to stay at home?), will allow the computer to elaborate on its own responses and disclose large amounts of information. Thus students are trained in recognizing and focussing on topics in steering a conversation, and in using culturally appropriate registers and forms of politeness in discourse.

Also in preparation are document-investigation exercises that allow students to explore a piece of "realia," in this case a page of apartment ads from Paris. Students will choose among these apartments, while learning the special vocabulary and abbreviations of apartment listings. Gilberte Furstenberg, lecturer in French, is preparing this exercise, as well as materials for interactive video, and is linking it to interactive video-based slides of the neighborhoods listed. The form of the document-investigation program is linked to a more general goal of ours, which is the creation of a simulation in the target language. Student participation will be through exploration of documents and through assumption of a role within the fictional world of the simulation. Students will be able to choose their mode of operation (invisible, visible, or role-assumption mode), and hence the degree to which they wish to interact with the fictional characters in culturally appropriate ways. The simulation will involve classroom activity as well as computer time. The work on the simulation is being done by Douglas Morgenstern, lecturer in Spanish.

Pedagogic Parameters

The materials we are developing are based on the sociolinguistic theory, established by such linguists as Vygotsky (1962), Widdowson (1978), and Candlin (1980), that linguistic competence, i.e., the ability to use correctly the grammatical and lexical structures of a foreign language, is a subset of a more general discourse competence that includes the ability to express, interpret, and negotiate meanings within the social context of interpersonal interactions. Indeed the ability to manage discourse in an interactional context not only facilitates the acquisition of the foreign language structures, it is the only way in which students can learn how to use them in socially and culturally appropriate ways.

Discourse competence can only be developed within a syllabus that stresses the process of learning rather than a product to be tested and measured; a syllabus that systematically trains students to recognize and interpret critically the intentions, goals, and beliefs of interlocutors in spoken discourse and of an author behind a written text; a syllabus that explicitly makes the student reflect not only on the grammatical forms of the language but also on the personal, social, and cultural communication strategies of language in use.

Our approach to the computer medium is thus essentially different from the approach that characterizes most existing computer-aided instruction. It is based on a philosophy compatible with Seymour Papert's philosophy of mathematics teaching. Applying what Papert says about learning math in schools to learning a foreign language in a classroom situation, we can say that classroom language learning is as different from language learning in real world interaction as school mathematics is different from the science called mathematics. The classroom's exclusive emphasis on the formal aspects of language has created the kind of "dissociated learning" Papert deplores in his Mindstorms (1983). Functional syllabi have not changed the model of learning a foreign language, just as the "new" math has not really changed the approach to school math. Language learning has remained positivistic, quantitative, "measurable."

By contrast, the materials we are developing -- called LINCS (Language-Learning through INteractive Communication-based Software) -- enable a student to get to know a foreign language the way one gets to know a person: by an interactional process of negotiation of intentions, goals, and beliefs, by epistemological reflection on what language can and cannot do, by knowing why the language operates the way it does, and not only that it operates in this way or how it operates.

The activities in progress are based on the following three principles:

1. They are dialogue-based. They all include a message to get across, an information gap to be filled between the computer and one or several students, a communicative task to achieve.

2. They activate the students' initiative. In order to have control over a conversation, students need to develop initiative in taking turns, initiative in reacting to what others have said, initiative in directing the topic of the conversation. The goal is less to learn or to react to information provided by the computer as teacher, than to find out what information the computer knows, or so to speak, to "teach the computer."

3. They are immersed in the foreign cultural perspective. If learning a language is acquiring a new set of tools to express the world around us, it is essential that the learners operate within the reality created by those tools, without reference to the mother tongue or the mother culture. The activities we are designing short-circuit the mother tongue and have the students build pictures, topics, and narrations totally within the micro world of the foreign structures. They activate the abilities of the learner to associate, compare, contrast, and group concepts and events according to the semantic networks called forth by the words themselves.

The computer serves several functions here, all of which are in direct opposition to the current view of the computer as tutor, drillmaster, prestructured provider of information, eternally patient error-corrector: a view that infantilizes the student rather than opening up new ways of experiencing and learning language. Among the more appropriate functions for the computer in the learning environment are the following:

1. It offers the multi-media, "discovery-rich" context of language in use, for a type of learning that is exploratory and "syntonic" (in Papert's phrase), i.e., coherent with and related to all aspects of the learner's worlds. Examples of this kind of learning are the microworld created by the simulation exercises and the multi-media environment created by interactive video systems that combine video, audio, text, and graphics.

2. It enables the learner to create and function within his or her own microworld with the foreign language, reconstructing as it were the microworld created between two conversational partners. Both ToPicks and LINGO work in this way because they are structured as conversations.

3. The learner can teach the computer, i.e., make the computer do things. In ToPicks, the learner teaches new words to the computer; in the simulation game the learner both elicits and enters information; in Reverse Eliza, the learner causes the computer to tell a story; in LINGO, the learner causes the computer to clean up or mess up a room.

4. It can oblige the learner to solicit input from the computer, finding a new way to do this with words, and reflecting on why the computer responds the way it does. This process is the same as learning to interact in the foreign language in natural settings, where you have to deal with _interlocutors_, not teachers. This will be the basis of the Tacit-Turn exercise and of many of the interactions within the Simulation.

5. It can slow down the construction of this microworld to permit reflection on the _process_ of discourse, i.e., to gain procedural knowledge. For instance, one can get grammatical cues in LINGO and discourse cues in Tacit-Turn.

6. It can develop a view of error correction as real-life interactional adjustment or "debugging," rather than as a right or wrong, true or false behavior, typical of school language. Our error correction will rest on echoing, expressions of impatience, and on an effort to understand the intent of the speaker. We can do this because of the artificial intelligence system at the heart of our exercises.

7. It can bring together teacher and learner into a common and joint exploration of language in use, in which the teacher does not know the solution, but goes through the process of solving the communication task together with the learners. For instance, there will be no "right" or predictable answer for learners who use ToPicks or who participate in the Simulation Exercise. Everyone who uses these tools will be a participant in inventing the environment.

8. It can revalidate language as a source of esthetic pleasure that stems from discovering the richness of meaning in words, and the possibilities of variation within the predictable structure of the language.

Answers to Some Concerns

This positive view of the uses of an advanced computer-based environment for language teaching does not ignore, however, some justified concerns over the appropriateness of the new technology. We list them below, and discuss them briefly.

1. Isn't there a paradox in using a computer to teach students real interaction with people in a foreign language? Not any more than using "teacher talk" to teach natural discourse. As compared to teacher talk, the computer offers a non-instructional, learner-monitored type of discourse AND the opportunity to reflect upon processes of discourse.

2. How can you teach spoken interaction by typing in written language? This is an advantage at beginning levels where vocabulary is limited and where most of the foreign language learned is in the written form. We are hoping to develop our capability for speech recognition in order to offer more speech access at later levels.

3. Speaking of user-friendliness, won't the learner be burdened with a lot of typing? No, we think there will be much more thinking than typing going on in the actual experience of the exercises.

4. How can students experiment with the language when there is an established norm of grammatical accuracy and cultural appropriateness? First of all, there are more variations here than the textbooks admit. Secondly, learners will always approximate their interlanguage according to their imperfect understanding of the language: rather than stifling this hypothesis-testing and experimentation process, we propose to use it!

5. By "computerizing discourse," aren't we rendering mechanical something that has the flexibility of fluidity of interpersonal relations? No, for two reasons: every-day talk is much more ritualistic than many people think; the LINCS materials make students aware of different styles of talk, some ritualistic, some creative. The computer encourages the acquisition of a metaknowledge of discourse.

6. Is metaknowledge of a language useful for learning how to speak it? Is not automatic reflex a better way of learning languages, as in the audio-lingual and immersion approaches? Even children "learn more and more quickly by taking conscious control of the learning process, articulating and analyzing their behavior," (Papert, p. 53). After all, the grammar-driven learning heightened the grammatical awareness of learners but did not enable them to interact with people; discourse-centered learning should give them a similar awareness of interactional processes and offer them the opportunity to act upon this awareness. In the end the knowledge transfer will only happen if the learner wants it and finds it relevant to his/her needs.

Conclusion

We are privileged to have access at MIT to advanced technology and world-class expertise. We are attempting to use this opportunity to create tools that reflect the most enlightened teaching methods and that make of the computer a new medium for the communication of meaning in words. The addition of video materials that will present multiple speakers and culturally rich materials in non-didactic, non-stereotypical situations, and of an interactive audio system that will allow a learner to master the pronunciation and intonation patterns of native speakers, will contribute to the creation of a uniquely powerful environment for language learning. The project includes the design of follow-up activities stressing the cultural and communicative function of language in the classroom. The goal is not to replace the teacher-student interaction, but to empower the student to extend the realm of learning.

This system has special relevance for the learner of Japanese and for Japanese and other Asian learners of English. The pronunciation/intonation practice system will at first be geared for ESL training of our many Asian-background students. Furthermore, the communicative approach to CAI advocated here is the most appropriate way of modeling behavior likely to achieve more than a superficial contact between our very different cultures. Although this approach might be considered as quintessentially western in its orientation toward freedom of exploration and negotiation of meanings, Japanese speakers can also benefit from learning how to negotiate verbal exchanges and extend conversations beyond the functional minimalism usually taught to businessmen. Although we are not including Japanese in the languages under development now, it will be relatively easy to develop such a system based upon the tools now in production, and we look forward to this possibility as a natural extension of our work.

REFERENCES

Breen, M.P. and C.N. Candlin (1980). "The Essentials of a Communicative Curriculum in Language Teaching," _Applied Linguistics 1_, pp. 90-112.

Papert, S. (1983). _Mindstorms: Children, Computers, and Powerful Ideas_, New York: Basic Books.

Vygotsky, L.S. (1962). _Thought and Language_, Cambridge: MIT Press.

Widdowson, H.G. (1978). _Teaching Language as Communication_, London: Oxford University Press.

TEACHING JAPANESE ON THE PLATO COMPUTER-BASED EDUCATION SYSTEM

J. Marshall Unger

PLATO is a time-sharing computer system with optional stand-alone terminal operation capabilities designed specifically for instructional purposes at the Computer-based Education Research Laboratory of the University of Illinois (Champaign-Urbana). Although one sometimes comes across such ex post facto decipherments as Program Logic for Automated Teaching Operations, PLATO was not intended to be an acronym; it is simply a registered trademark of the Control Data Corporation, which manufactures the CYBER mainframe computers on which PLATO operates.

PLATO is particularly well-suited to the task of teaching scientific and technical Japanese. It supports an outstanding library of lessons (computer-assisted instruction programs) in chemistry, physics, computer science, and other scientific disciplines; and, as the result of work by me and my assistants in Honolulu, it is now equipped with Japanese script-handling features. These features are fully integrated into PLATO system-level software, which provides a robust foundation for the development of instructional materials, and have been extended to off-line PLATO (Micro-PLATO) by programmers at Control Data Japan.

Although my involvement with PLATO in Hawaii began in 1977, the University of Hawaii did not acquire its own PLATO system until December 1984. Throughout the previous years, PLATO services were provided by a telephone link to the University of Illinois, and it was never certain if or when the University of Hawaii might "pull the plug." In addition, during this time the number of terminals available for student use was very limited. An average of 1,100 undergraduates take Japanese-language courses at the Manoa campus (the largest of eight) every semester; perhaps three or four hundred of these are in elementary language courses, and since PLATO services were also being used extensively by courses in physics, agricultural biochemistry, dental hygiene and so forth, it was impossible to make PLATO a compulsory part of the language program.

Finally, in producing Japanese lessons for PLATO, my colleagues and I were concerned with the unique motivational and cross-cultural problems that the extremely heterogeneous student body at the University of Hawaii poses. For these three reasons, we concentrated for the most part on the production of supplementary, textbook-independent lessons at the elementary level, which explored the full potential of PLATO software and steered away from a more comprehensive curriculum-oriented approach to lesson development. My comments about the use of PLATO for the specialized job of teaching scientific and technical Japanese will therefore be mostly prospective. I will begin with some observations about computer-based education in general, illustrate some key PLATO features with a simple drill-and-practice example, and conclude with some suggestions for using PLATO to help specialists learn Japanese.

Consider the situation of a student using a computer terminal to deal with multiple-choice questions presented by a bare-bones CAI program. At this zero-point, questions and answers are completely fixed beforehand. So are the various feedback messages to the student's responses and the contingent program branching that depends on those responses. As a whole, the question-and-answer modality of the situation is fixed: the student cannot switch freely to some other kind of CBE activity. In these three ways, a simple multiple-choice program is highly formalized and constrained, hardly more than a translation of programmed instruction from a paper-and-ink workbook to an electronic medium. Pedagogically, too, it is a zero-point. To the extent that the student is forced to choose among multiple answers, he or she is liable to be misled rather than enlightened. To the extent that the program is unable to alter the order and content of the material presented, it fails to emulate a real teacher. And to the extent that the student/machine relationship is confined to a particular type of activity, the student's sense of active participation, personal responsibility and motivation are diminished. In short, it is a waste of technologically sophisticated resources to use computers in narrow accord with this zero-point model.

There are basically three directions in which we can move away from this zero-point into more productive CBE usage. By replacing the multiple-choice format with routines that collect ordinary natural-language input from the student, we can increase the student's freedom of action. This encourages free expression within reasonable limits, eliminates the need to display misleading "answers," and relieves the monotony of menu-style input (which, of course, would still be used for many operations ancillary to the main business of learning the subject matter at hand). Similarly, by replacing strictly predetermined branching, feedback, and question selection with dynamic program actions, we can give the system a more human feel. Not only can both exercises and feedback be modified at random or in accordance with the student's past performance, they can also be made to incorporate variable data supplied by the student. Finally, if we replace the concept of "teaching machine" with the broader concept of "learning environment," many easily overlooked opportunities to improve the overall quality of the man/machine interaction become readily apparent. Tutorials and drills can be supplemented with simulations of laboratory experiments and other real-life situations. One can set up interactions in which the student asks the questions and the computer answers them -- the direct opposite of the classical CAI arrangement. On a network, real-time communications, electronic mail, multi-user forums, polls and so forth can be brought into play.

In my opinion, the design of CBE systems is essentially a matter of maximizing the distance from the zero-point of this model along all three dimensions: increasing student freedom, reducing program determinancy, and improving the quality of the man/machine interaction as a whole. Physical constraints, particularly with stand-alone microcomputers, such as limited memory capacity and long disk-access time, may require some compromises; but in general, "thou shalt not" limit the student's freedom of action, the program's flexibility, or the number of functions the system as a whole can provide without some compelling reason.

Figures 1 through 5 show a somewhat whimsical example of these three "commandments" in action in one module of a Japanese lesson that gives students practice with sentence-modifier patterns, the homologs of relative clauses in English. As the Japanese sentences at the top of the display indicate, only the creature on the right has a navel; the question below asks, "Which frog is the real one?" In Figure 2, the student has typed in, "It's the one on the right," and the lesson has given its feedback. Figure 3, "It's the one on the left," is another perfectly grammatical answer -- in this case, factually correct as well -- but the lesson is looking for a more complete answer. Figure 4 shows a nearly perfect response; the lesson has spotted a grammatical error this time. Finally, with Figure 5, the student has produced one of the many possible correct answers that contain a sentence-modifier. Although very simple, this example illustrates two key points, one of importance for CBE in general, and the other specifically related to teaching Japanese on computers.

The general point is that the more natural the interaction is from the student's viewpoint, the more complex and extensive the underlying software must be. (I cannot resist drawing an analogy with the performance of music, in which technical discipline is the prerequisite for expressive freedom, and what sounds spontaneous is in fact carefully practiced.) In this example, phrase-structure rules coupled with vocabulary lists are used to judge student input. The lesson author can optionally keep a detailed record of this input for later analysis. Minor spelling errors and variations in Romanization are tolerated, and appropriate feedback to hundreds of different sentences can be provided in a few seconds. Answers are not judged on a yes/no basis, but rather in accordance with the three-way branching "ok," "wrong" (anticipated error), and "no" (unanticipated error). Instead of having to type a new answer on each try, the student can edit the previous answer, request help at any time, leave a note for the lesson author, attempt to page the instructor, and so forth. Leaving aside the network features, the programming involved to provide these features on a microcomputer using, say, BASIC, would be prodigious; the TUTOR language used on PLATO makes it easy, allowing the author to concentrate on the broader logic of the interaction instead of getting bogged down in the details of programming.

The Japanese-specific point illustrated by this example is that students do not need to use native script for input in instructional lessons except in those cases where the correct selection of script symbols is itself the object of the exercise. Indeed, in order to judge the validity of a response containing Chinese characters (<u>kanji</u>), it must either be converted into phonemic script, or else it must be matched, code number by code number, against a template, with no room for alternative spellings. In the first case, the work that went into selecting <u>kanji</u> is undone and therefore counterproductive; in the second case, we are back to multiple-choicing, in disguised form. In both cases, the effect is merely cosmetic, the student has been distracted from something more important, and the possibility of misleading feedback (a correct answer judged wrong because of one or more unanticipated <u>kanji</u>) has been needlessly introduced. The exclusive use of Japanese syllabic script (<u>kana</u>) also leads to problems unless spaces are used to separate words and phrases, which is not customary in ordinary Japanese orthography. (For an extremely thoughtful discussion of the whole problem of Japanese input on computers, see Yamada 1984.)

There is a link betwen these two points worth noting. The more complex the software, the greater the temptation to dazzle the end-user with a show of programming skill and system features; this temptation can feed the already well-developed obsession with native script that many teachers of Japanese share. A CBE author of Japanese material should therefore be doubly cautious. To give a concrete example, I have developed a technique in TUTOR for letting the student type whole-sentence responses entirely in <u>kana</u>, without spaces, and parsing them very quickly so that variations in the order of words and phrases can be taken into account during the response-judging process much as it is in the case of English responses. I would be the first, however, to discourage anyone from using this minor masterpiece of hacking in a serious PLATO lesson aimed at beginning students of Japanese. Just because something can be done on a computer does not mean that it must have some immediate educational application. Unfortunately, this seemingly obvious proposition enjoys far less than universal acceptance.

At a recent symposium, one group of researchers announced a Japanese CBE system aimed at foreigners that "...teaches Japanese by teaching <u>kanji</u>" (Connell, Nagamatsu, Takeda & Ohiwa 1984:35-37). Their method is built around a so-called touch-typing system devised by Ohiwa, and a texbook based on a <u>kanji</u> dictionary by Nagamatsu. Each lesson begins by drilling the two-stroke input codes of a few new <u>kanji</u>. This is followed by drills for inflected forms (i.e., <u>okurigana</u> usage), whole sentences, grammar, the proper reading of <u>kanji</u> in context, and translation; some lessons have no grammar drill.

In all these, interaction with the computer involves the use of Ohiwa's J-TYPE input codes. Despite this heavy emphasis on _kanji_, the authors are convinced they are teaching the Japanese language _in toto_, as is shown by the comparison of their system with "many of the intensive language courses now in operation in Japan," and the fact that the only prerequisite they mention is "an adequate ability in English."

As a linguist and language teacher, I find this _kanji_-oriented approach and cavalier disregard for the preeminent role of syntax in elementary language learning truly appalling. It is a classic example of what John De Francis (1984) has aptly called the "Indispensability Myth" of Chinese characters. Even for teaching students beyond the elementary level, my experience with PLATO leads me to conclude that the right way to build a computer system for helping students learn Japanese is to start with a good CBE system and add Japanese script-handling capabilities, not with a Japanese word-processor and add educational enhancements. I think the following examples bear this out.

I have already made a case for Romanization in the early stages of Japanese language learning, but, of course, the study of _kanji_ is unavoidable; indeed, were it not for _kanji_, I doubt that Japanese would be popularly regarded as such a formidably difficult language. Because of PLATO's high-resolution displays, fast plotting speed, and large memory capacity, my assistants and I designed _kanji_ and other Japanese graphics using vectors rather than dot-matrix patterns. Figure 6 shows some of these. By encoding _kanji_ in vector format, a considerable amount of memory was saved, and the same characters can be displayed in different scalings, rotated, and even in mirror image. More importantly, vectors make it possible to record all the stroke-order and stroke-direction information for each symbol while, at the same time, maintaining typographically correct standards. Note, in particular, that the doubling-back in the third stroke and the break in the sixth stroke do not cause a false stroke count. Thus, the PLATO Japanese graphics data base can be used to teach and review writing skills. This is not a top-priority area for technical language learning, but it definitely is one for which the conventional classroom setting is inefficient, since each student needs individual oversight; hence it is an ideal target for computer-based education.

Incidentally, Figure 6 is taken from a lesson that could be expanded into a complete on-line Chinese-Japanese character dictionary (_kanwa jiten_). The student can ask to see all the _kanji_ from the Japan Industrial Standard Level-1 set of 2,965 which take a certain reading or contain a certain radical; if a radical is specified, so may be the number of residual strokes in the characters. The program then quickly lists all characters that fit the specified description. Note that the combination of reading and radical look-up _simultaneously_ is not possible with a standard hardcopy dictionary. This kind of information retrieval capability has obvious applications in computer-assisted translation exercises.

Perhaps the most exciting possibility for foreign-language CBE, however, lies in the area of simulations. Figures 7 through 10 show part of an off-line PLATO lesson used for Boeing 747 cockpit training. Figure 7 is the engine start-up/shut-down checklist; the object is to carry out the steps listed by touching the switches and buttons on the display screen shown in Figure 8 in the correct order. I have managed to do this by Figure 9, and am ready to try engine shut-down. As Figure 10 indicates, I made many mistakes before completing the task, and can repeat it as often as necessary until I get it right. Material of this kind could be turned into a reading comprehension lesson for aeronautics engineers without much trouble, thanks to PLATO's courseware development editors. Of course, since the object would no longer be the operation of the airplane, branching through the lesson would be somewhat different. Supplementary questions would probably be interpolated, various kinds of tutorial help (Romanizations, translations, etc.) would be available, and so forth.

Figures 11 through 15 show a less technologically demanding simulation with perhaps greater potential for language teaching purposes. By touching a screen, the student simulates the laboratory distillation of a mixture of water and alcohol. In order to keep the number of figures small, I have omitted those showing mistakes and corrective feedback, although it is primarily through errors that the student learns in this type of exercise. Again, it should be clear how this material could be reworked for reading comprehension. Since chemistry students already know the English names for the various pieces of apparatus, we could build a separate vocabulary module around Figure 11; alternatively, we could have students request the various pieces of apparatus in Japanese. In either case, the assembly task shown in Figures 12 through 14 would serve as a review. The situation shown in Figure 15 could then be used as an opportunity for students to give orders and ask questions in Japanese; this would require removing some timing loops from the original lesson so that they still would not blow up on students who made mistakes, but from a programming standpoint, that would actually be a simplification.

Finally, let me mention the possibility of allowing students to use the Japanese word-processing features that were developed at the University of Hawaii, details of which are given in the appendix below. In principle, these features can be accessed by any kind of system user -- student, instructor, or lesson author -- at any point at which a response is requested. Thus, lesson displays, responses to questions, notes to other users, rosters, curriculum catalogs, and so on can, if desired, be handled in ordinary Japanese script. In practice, however, it is questionable whether even the most advanced students would profit from the experience of, say, writing compositions in Japanese script on PLATO. There is no easy way for the instructor to mark up a composition of this kind, and unless the student is going to correct the original, the whole exercise is better done "the old-fashioned way" with pencil and paper. In fact, the passive nature of *kanji* selection on a computer may be counterproductive for the mastery of Japanese writing.

On the other hand, as already remarked, PLATO software has not been built around a Japanese word-processor. Japanese script-handling features have been built into a pre-existing matrix of PLATO software; therefore, if someone were to come up with a new pedagogical strategy in which the use of Japanese word-processing software can play a constructive role, it would be easy to implement and test on PLATO. Word-processing thus brings into sharp focus the need for a balanced approach to the use of computers in Japanese-language training; it is not fancy graphics or synthesized sound or so-called artificial intelligence that make CBE distinctive, but rather the use of computers as electronic environments for all aspects of the educational enterprise -- instructional, administrative, and collegial. If work on PLATO does not yet speak directly to the problem of teaching technical and scientific Japanese, it nonetheless illustrates this point admirably, and therefore can serve as a guide to future developments in the field.

REFERENCES

Connell, G., Nagamatsu, M., Takeda, N., Ohiwa, H. 1984. "J-TUTOR: A Kanji-oriented Japanese Language Education System." 'Shizen Gengo Shori Kijutsu' Shinpojiumu (Symposium on "Natural Language Processing Technology"), pp. 35-41. Joho Shori Gakkai.

De Francis, John. 1984. The Chinese Language: Fact and Fantasy. Honolulu: University of Hawaii Press.

Yamada Hisao, "Wapuro to Nihongo no genjo to shorai" ("Word processing and the present and future of the Japanese language"), Nihongogaku 3:7 (July 1984) pp. 4-17.

Figure 1

Kono kaeru wa heso ga arimasen.

Kono kaeru wa heso de tya o wakasite imasu.

Dotira no kaeru ga hontoo no kaeru desu ka?

>

Figure 2

Kono kaeru wa heso ga arimasen.

Kono kaeru wa heso de tya o wakasite imasu.

Dotira no kaeru ga hontoo no kaeru desu ka?

> migi no desu *sigh*

Frogs are amphibians!

Figure 3

Kono kaeru wa heso ga arimasen.

Kono kaeru wa heso de tya o wakasite imasu.

Dotira no kaeru ga hontoo no kaeru desu ka?

> hidari no desu.   naruhodo

But how do you know that that one is the real one?

Figure 4

Kono kaeru wa heso ga arimasen.

Kono kaeru wa heso de tya o wakasite imasu.

Dotira no kaeru ga hontoo no kaeru desu ka?

> heso ga nai kaeru wa hontoo no desu.   iie

wa won't do: use *ga* as the question does.

Figure 5

Kono kaeru wa heso ga arimasen.

Kono kaeru wa heso de tya o wakasite imasu.

Dotira no kaeru ga hontoo no kaeru desu ka?

Heso ga nai kaeru ga hontoo no kaeru desu.   hai

The frog *which has no belly-button* is the real frog.

Figure 6

Japan Industrial Standard #4208

Stroke # 8

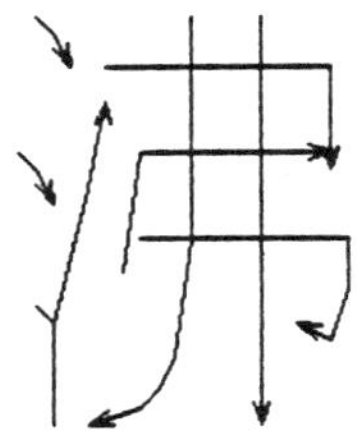

NEXT to step through

TOUCH a kanji to display its stroke order diagram;
or press BACK for another selection.

洲 沫 波 況 没 決 池 氾 水
浄 油 泊 治 沃 沙 汝 汚 永
浅 洩 泌 沼 泳 汰 汎 汗 氷
洗 海 沸 泰 沿 沢 沖 汲 求
津 活 法 注 河 沈 汽 江 汁
　 洪 泡 泥 泣 沌 沓 汐 汀

Figure 7

オペレーション

パネルの始動準備

バッテリー・スイッチ..................ON
DC メーター・セレクター.........APU BATT
APU 火災警告およびスクイブ.........CHECK

APUの始動

APU マスター・スイッチ...............ON
BLEED AIR スイッチ................CLOSE
APU マスター・スイッチ...........START

APUが正常に動作している時
BLEED AIR スイッチ.................OPEN

APUを停止

APU マスター・スイッチ..............OFF
DC メーター・セレクター............BATT
バッテリー・スイッチ.................OFF

画面をタッチしてください

Figure 8

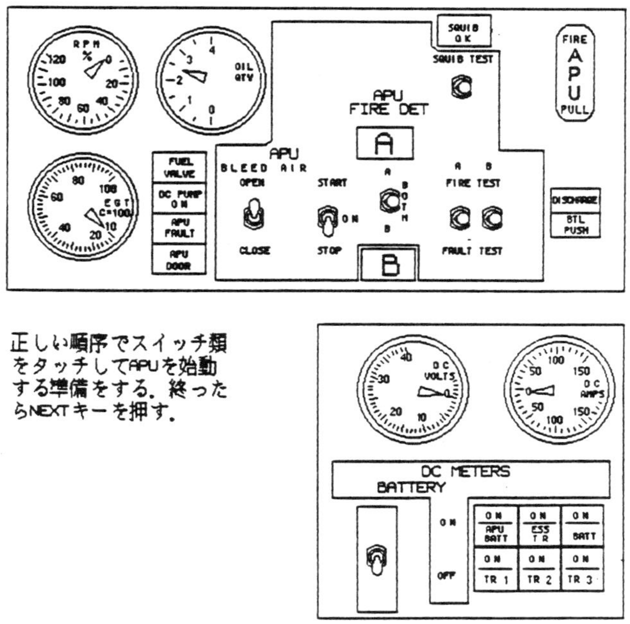

Figure 9

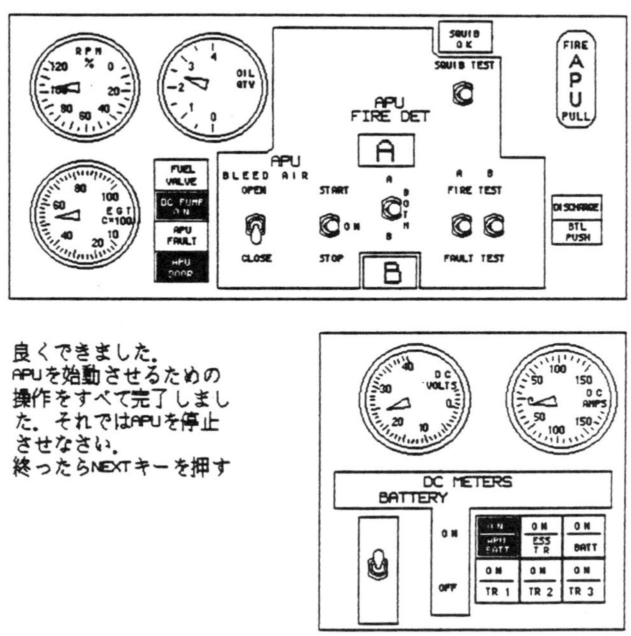

Figure 10

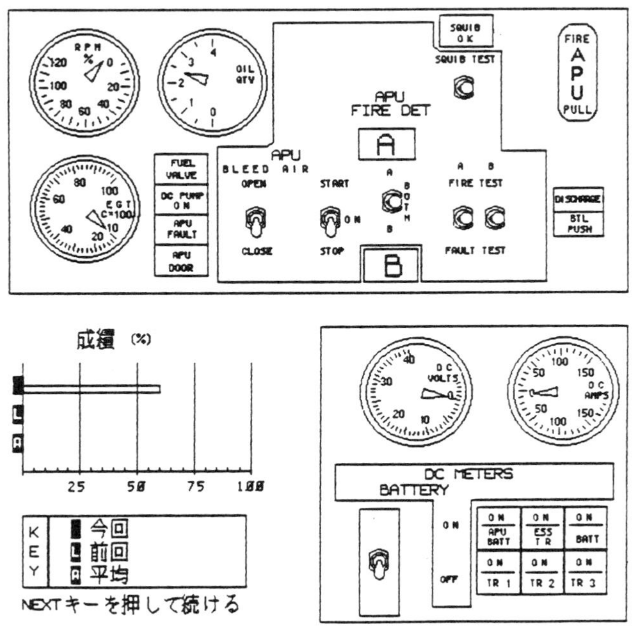

Figure 11

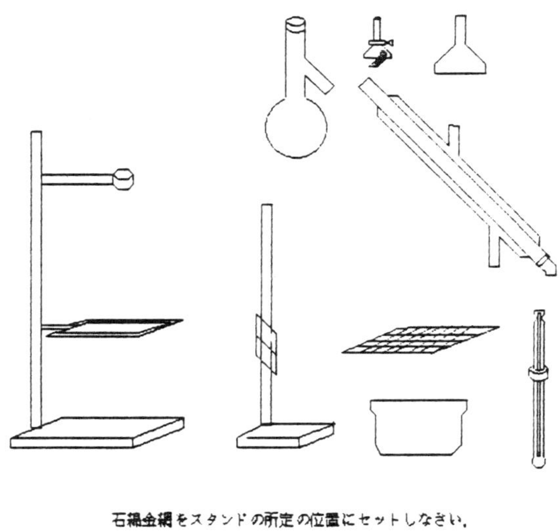

Figure 12

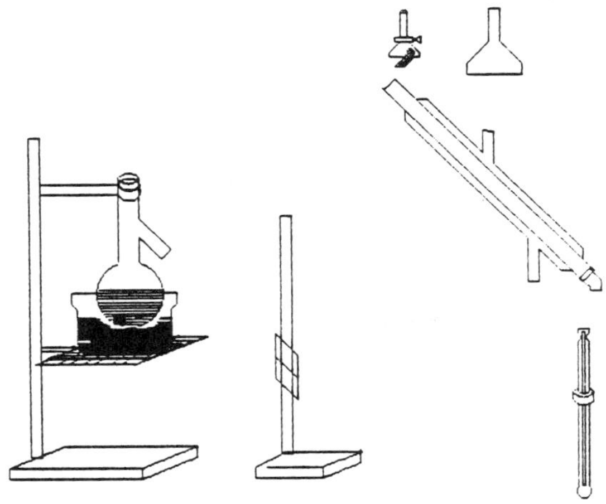

Figure 13

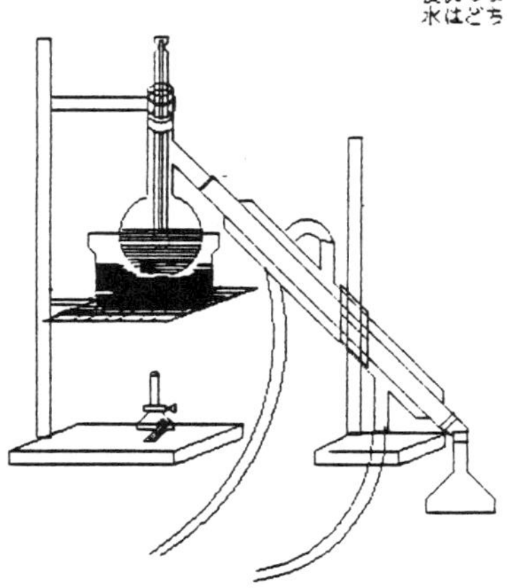

Figure 14

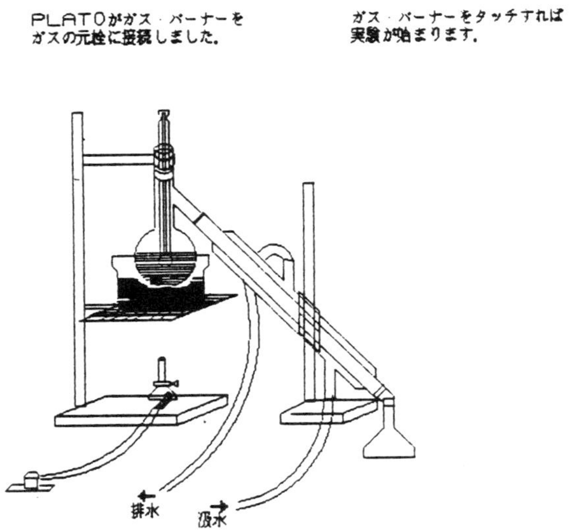

Figure 15

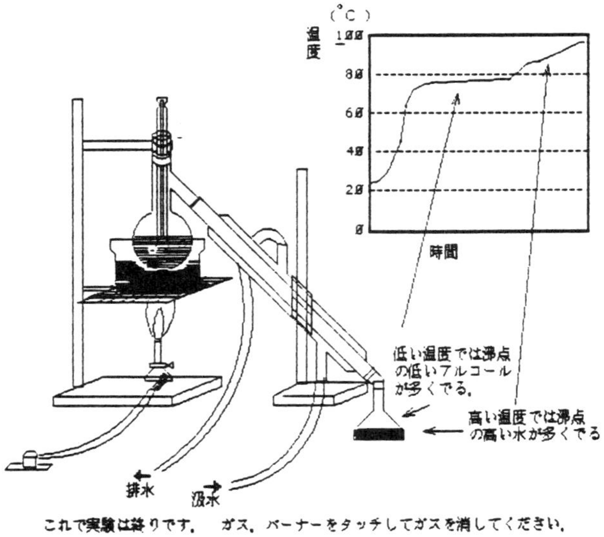

Technical Summary of Implementation of Japanese Script on PLATO

There is no "handshaking" between a PLATO terminal and the mainframe, and no local echoing of keypresses. Instead, all keypresses are transmitted to the host mainframe; at the end of each user's time-slice, the host formatter sends graphics directives to the terminal, where they are carried out by a local microprocessor. Both high-speed vector and dotmatrix graphics are available. In the case of line-drawn (vector) graphics, the formatter generates absolute screen coordinates; local firmware calculates the positions of intermediate pixels. Character graphics are handled dfferently: an alternate font memory area in the terminal is first loaded from the host with up to 126 patterns of xteen-by-eight pixels each; these patterns are then accessed in the same way as the standard character set. Because the telephone communication link between the host and the terminal has a limited transmission speed, the terminal firmware can draw lines considerably faster than coordinates can be passed to it, and it takes nearly seventeen seconds to load a complete set of alternate font characters. Thus, in order to plot extended passages of Japanese as fast as or faster than a user can read them, a peripheral storage device must be used -- currently this is a floppy disk. In both cases, an auxiliary terminal resident program is downloaded from the host or brought in from the local storage medium to handle character addresses when they are received from the host. The diagram below shows how the system is configured:

Components of the On-Line Japanese PLATO System

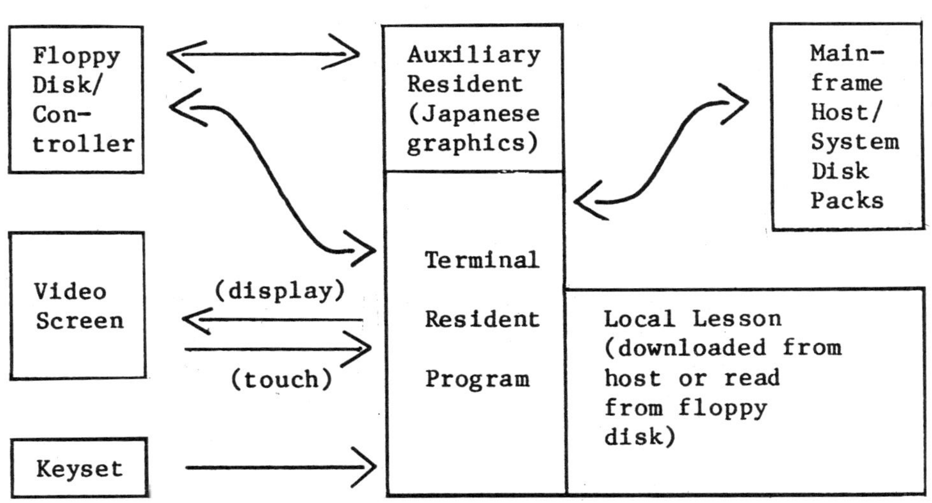

At the host level, PLATO uses 60-bit words and 6-bit bytes. For the purpose of representing Chinese characters and other graphics internally, the sixty-three possible bytes are divided into two sets: octal 1 through octal 40 (the charactes "a" through "z" and "0" through "5" are used to form base-32 numbers which address the characters; the remaining octal values are used as delimiters. Escape codes mark Japanese data strings, and indicate to the formatter whether characters are to be plotted horizontally or vertically. One such code is modified by the program which creates semi-compiled or *condensed* code from source code written in the TUTOR language; thus, authors can input strings in horizontal format that students will later see in vertical format. The formatter interprets Japanese data strings as ordinary alphanumeric characters or local-address data depending on the value of a flag in the user's station bank which can be set either from the keyset or by a TUTOR command. Thus, users may, if they wish, input Chinese characters, *kana*, and so forth as code groups of one to three letters similar to English words (see Figure 16 for examples). These codewords are derived from the Japan Industrial Standard transmission codes; currently, the 2,965 JIS Level-1 *kanji* are available. In practice, this mode of input would be used only by specially trained personnel for large-volume transcription.

For most purposes, users would input Japanese interactively. PLATO terminals are equipped with a TERM key that can, in principle, be active all the time, even when the system is waiting for user input at a standard prompt, called an *arrow*. When the TERM key is pressed, operation of the current program is suspended, the status of the user's station bank is saved, and the message "What term?" appears near the bottom of the screen accompanied by an arrow. The user then types the name of a subroutine, and presses the key labelled NEXT to begin using it. System-supported subroutines include "spell" (an English dictionary), "calc" (an on-line calculator), "talk" (page other users), and "comments" (leave a note for the author of the current lesson); user programs can include their own subroutines. Both system and user-defined TERMs can modify the screen display and make free use of arrows. Upon exiting from a TERM, the original station status is restored, and, if the user was waiting at an arrow, it is ready to receive input at once; the screen display retains the effects of the subroutine.

User-defined TERMs cannot pass data back to the buffer of a suspended arrow. For the sake of handling Japanese, however, this is now permitted for certain system TERMs. So far, only one such TERM has been defined, TERM-J, a roman-to-Japanese conversion routine. In the future, this will probably be upgraded, and a *pinyin*-to-Chinese routine may be added. In fact, different kinds of editors for the same language could coexist on the same system since the specifics of the dialog initiated by TERM are completely open. Any input method that can be programmed in the TUTOR language can thus be incorporated into PLATO: large keysets, tablets and digitizers can be attached to the terminal and interfaced with the mainframe in this way; conversion algorithms based on character readings, radicals, stroke counts, or any other properties of Chinese characters can similarly be handled. For that matter, the same technique could be adapted to entering two-dimensionally formatted mathematical expressions or musical notation at arrows if there were a need to do so.

Another advantage of this approach to Chinese-character input is that it works at all arrows regardless where they occur. This means that all types of users -- authors creating lessons, students using them, and instructors managing student groups -- are treated uniformly whenever they input Japanese. It makes no difference whether one is entering text in a notefile, writing TUTOR code, or using a display editor. Moreover, since standard alphanumerics and Chinese-character data can be intermixed, all the response-judging software provided by the TUTOR language will work for Chinese character input just as it does for English.

PLATO terminals can operate without a connection to a central mainframe. Lessons for stand-alone use are written in a language called Micro-TUTOR using central PLATO utilities; condensed (semi-compiled) code is then downloaded to floppy disk. Authoring in Micro-TUTOR is not possible locally; but by using such Micro-TUTOR packages as control Data's PCD1, one can develop and deliver lessons entirely off-line. PCD1 has two parts. In author mode, a user creates displays, sets up response judging, and in general specifies how and when to carry out all the actions that are normally associated with a Micro-TUTOR -arrow- command. Instead of generating Micro-TUTOR code, however, PCD1 keeps track of the user's decisions in a dataset stored on floppy disk. This dataset, when accessed in student mode, is interpreted by a driver which, in effect, executes a Micro-TUTOR lesson.

Thanks to PCD1, it is possible to author Japanese lessons off-line. Japanese input is currently handled by a large keyset ("table") attached to the external port of the terminal. Unlike the case of on-line PLATO, the formatter (now in the local microprocessor) plays no role. When PCD1 receives a character identifier, it passes this value to a Micro-TUTOR procedure which, in turn, calls a machine-language subroutine that retrieves and plots the appropriate character. Thus, for off-line Japanese, the terminal resident program and the Micro-TUTOR interpreter undergo no special modifications. This guarantees that Japanese PLATO will be able to use future versions of off-line PLATO software without major adjustment.

ENHANCED LANGUAGE INSTRUCTION:
INTERACTIVE VIDEO AND AUDIO TECHNOLOGY

Randall L. Jones

Even though computer-assisted language instruction (CALI) has existed in some form for at least 15 years, the past two years have shown a significant increase in the interest and involvement of language teachers in this new technology. Not only the sudden appearance of many computer language learning programs, but also the number of publications, papers presented at professional meetings, new special interest groups, and even organizations dedicated to the computer and language teaching attest to the fact that language teachers are serious about the viability of the computer as a device for learning. As we allow ourselves to be dazzled by the capabilities of the computer as a tutor, it is important that we are also aware of its limitations.

The use of a computer in a language program is especially valuable for individualized learning. Concepts presented in the classroom are not comprehended with equal facility by all students. The computer provides the possibility for the student to control the rate of presentation, the amount of repetition, and the extent of illustration. Furthermore, because the student interacts with the computer, both physically as well as cognitively, the experience is usually very satisfying. On the other hand, the computer is quite limited as to the types of real language interaction it can provide. Text can be presented on the screen and written by the student via the keyboard. For this reason, CALI programs have been principally grammar and vocabulary tutorials, drills and games.

But with the use of additional existing technology, the computer is capable of far more than tutorials and drills on grammar and vocabulary. With the addition of interactive audio and video devices, the computer can serve as a medium for organizing and presenting virtually unlimited amounts of real language material and allowing the student to interact with it. I would like to illustrate one approach to the use of intractive video and audio in a foreign language learning environment by describing the enhancements to the Defense Language Institute's German Gateway program, which BYU and CALICO have been involved with for the past 18 months. I stress that this is but one application of this technology. I feel certain that creative minds will eventually design technology-based language learning strategies that go far beyond our simple procedures of today.

Let me begin by giving a brief description of the videodisc and its capabilities. On one side of one disc, it is possible to store up to 30 minutes of motion video or 54,000 still frames, or a combination of both. There are also two audio tracks and a special data track. An industrial model videodisc player can access any section of the disc in a few seconds or less, and play forward or reverse at regular speed, slow, or fast motion. Any of the individual 54,000 frames may be indefinitely displayed as a still frame, thus simulating a slide projector, without any wear to the disc or player. Under computer control, the material from the videodisc can alternate with textual information and graphic images generated by the computer. In addition it is possible to superimpose computer generated information over the video images.

For the German Gateway program, our task was to augment an existing German language program using conventional text and tape media with situationally-oriented material using interactive video. Our basic goal was twofold: to provide authentic context for the language which was being learned from the book and tapes; and to provide useful listening comprehension material as a basis for future oral production. No attempt was made to use an existing speech recognition device for spoken interaction between the student and the computer.

As hardware, we selected the SONY SMC-70 microcomputer with superimpose board, the SONY LDP-1000A videodisc player, and the SONY PVM-1270Q RGB and composite video monitor. The SONY configuration was actually mandated by the contract, but it was also the best computer-controlled videodisc system available at the time. We have recently added the EIS Instavox for additional interactive audio. The lessons were first written using the authoring language PILOT Plus, but we are now using a new SONY product called Courseware Development System. For video material, we used portions of the BBC German program, Kontakte, distributed in the United States by Films Inc. Because the Kontakte material does not include all situations required by Gateway, it was necessary to shoot some original footage on location in Germany.

The videodisc portion of Gateway centers around a series of language situations including getting a hotel room, ordering in a restaurant, asking information and purchasing a ticket in a train station, getting gas, etc., in a service station, having a car repaired, renting a car, renting an apartment, asking directions, riding on a streetcar, discussing politics, and being invited to someone's home. Each situation has several video segments to represent the variations possible for that situation.

As the student begins a computer session, he is asked to select from a menu the desired situation. Within the situation, he is then asked to select a segment. At this point, we use a four-prong approach. First the student sees the entire segment, usually lasting no more than two minutes. The segment is then replayed in smaller sections or subsegments, each being a self-contained language interchange. After each subsegment is played, the student is presented with three options: help, exercises, or a quiz. The quiz must be passed in order to proceed to the next subsegment. The exercises are designed to help the student use the newly learned language material. The help consists of three parts: see the subsegment again, see the subsegment again with the German text superimposed on the screen, and vocabulary assistance. The Defense Language Institute has requested that the lessons contain no grammar drills and that English not be used in any but the introductory lesson. The vocabulary assistance therefore consists of visual helps, either from the video material directly or generated by the computer graphics program. This does, of course, present severe limitations. It is simply not possible to find or create suitable visual images for many of the essential words in the lessons. Furthermore, computer generated graphic images require large amounts of space on the data disks. We decided, therefore, that students would need to have ready access to a good dictionary as they use the program.

After the student has worked his way through the subsegments, the segment is played once again in its entirety, so that the student can rebuild the whole language interchange from the information learned at a lower level. Finally, a test is given for the complete segment.

As an example of how this all works, in the lesson on hotels is a segment in a Verkehrsamt. A Verkehrsamt is an office located in or near the railway station for assisting travelers in finding a hotel room. The student sees a man walk into the office and ask about a room. The woman asks what kind of room he wants, to which he replies a single room with a shower. She makes two phone calls, but is successful only in finding a room without shower or bath. He agrees to take the room. She gets some information from him and explains how to get to the hotel. The entire interchange takes 2:02 minutes and contains 247 words, many of which are used several times. Within those 247 words are several basic phrases used for greetings, expressing gratitude, asking for information, etc.

After the student has seen the entire segment, he sees the first subsegment, which consists of greetings, the request on the part of the traveler, a question from the Verkehrsamt clerk, and the traveler's response: a total of 16 words and about six seconds. By selecting option 1 on the HELP menu, the student can see the subsegment again. The videodisc player is able to return to the beginning of the subsegment and begin playing in about one second. This very fast access speed is one of the important features of a videodisc-based interactive system. Although it is also possible for a computer to control a video cassette player, the response time is much greater, often causing loss of interest on the part of the student. Again the student has the option of repeating the subsegment as often as is desired.

By selecting the vocabulary help option, the student can receive assistance for those words that can reasonably be demonstrated using video or computer-generated graphic material. For example, the student can hear the word Einzelzimmer (single room) while at the same time seeing a simple illustration of a single bed on the screen. The same is possible for Dusche (shower) and Bad (bath).

The lessons are written in such a way that students are not encouraged to rush through a segment. We feel that repetition is important, and several cycles are necessary before the student comprehends the material at the desired level. Because there are usually several different segments of the same situation, students find each successive segment easier to understand. Although we do not intentionally write the material using gaming techniques, we have found in initial tests that students respond to it in much the same way that they respond to a video game.

Shortly after we completed some of the initial lesson material, it became obvious to us that an additional external audio device would be of great value. We already had recorded on the second audio track of the videodisc some special material such as questions and useful expressions, but when a videodisc is made, it cannot be changed. We wanted flexibility to add new material as needed. Furthermore, because a video still frame requires only 1/30th of a second, it is not possible to use the audio track. We wanted the possibility of showing individual frames and having audio to reinforce it.

At the current time, there are four types of interactive audio devices which can be controlled by a computer. The first is the interactive cassette player. It is much like a conventional cassette player in most respects, but it can connect with a computer so that commands in a program or lesson can operate it, i.e., fast forward and reverse to a location on the tape, play, and record. At the present time, the only such player is the Tandberg 812, but SONY plans to introduce a similar model sometime this year. We have found the Tandberg 812 to have relatively fast access time, but not fast enough for our purposes. When working with a videodisc, the comparatively slow response of a cassette player can be disturbing. It is, however, a marvelous machine and can be used well for many other types of language instruction.

The second type of computer-controllable audio device is the random access audio disk. It is much like a computer disk in its design and operation, but it contains audio material instead of digital data. The only device of this kind now on the market is the Instavox, manufactured by Educational & Information Systems Inc. Each 15-inch disk holds up to 30 minutes of high quality recorded material. The worst case access time is .4 seconds, i.e., faster in most cases than the videodisc. The Instavox is more expensive than the Tandberg 812, but the difference in price is more than offset by its speed.

The third possibility for a computer voice is digitized speech. The utterance is converted from an analog to a digital signal, and stored on a computer disk along with other data. It can be called from a program with instantaneous response. The quality depends on the sampling rate of the device, and can exceed the quality of magnetically recorded speech. Its greatest drawback is that it requires a great deal of disk space for utterances of even a few seconds. I feel, however, that it will become a standard item for foreign language CAI lessons in the near future.

Finally, there is the possibility of synthesized speech. It is not recorded, thus it does not require space on a disk. It is virtually unlimited in the number and types of utterances it can produce. Recent models, such as the DEC Talker, can also make variations for sex and age. The only real drawback of synthesized speech is that it still has not been perfected to the point that it really sounds like human speech. We need to be very careful about the kind of pronunciation models we provide to our students. Synthesized speech for language learning may be a reality some day, but for now it is still in the development stage.

We plan to have the majority of German Gateway lessons completed by the 15th of April, 1985. They will be used by the Gateway students at the DLI and evaluated thoroughly by the Army Research Institute at Monterey. We are encouraged by the initial testing that we have done. We are optimistic that the video material will indeed provide an authentic context for language learning. We are also certain that the students will find them a pleasant supplement to the method now being used. Our collective goal is to assist the students in learning the material more thoroughly and more efficiently.

Let me also mention that BYU and CALICO are working on three other interactive video projects for the Defense Department in Spanish, Korean, and Hebrew. In each case the approach is somewhat different. We are using different hardware and software for the Korean and Hebrew projects, and hope to be able to have valuable comparative data within a short time. There is a good chance that we will begin work in new languages sometime this year, possibly Arabic, Japanese and Russian.

There is no question that language learning can be greatly enhanced by the use of interactive video and audio. At the present time, the cost is somewhat prohibitive, and the technology is still not entirely stable. But that is changing. With the development of digitized video, the videodisc may soon be a relic of the past. The prospects are exciting and the applications are limited only by human imagination.

REFERENCES

Gale, Larrie. "Montevidisco: An Anecdotal History of Interactive Videodisc." CALICO Journal 1,1 (June 1983), pp. 42-46.

Higgins, John and Tim Johns. Computers in Language Learning. Reading, MA: Addison-Wesley, 1984.

Underwood, John H. Linguistics, Computers and the Language Teacher. Rowley, MA: Newbury House, 1984.

VI

CONFERENCE PARTICIPANTS

Mr. Robert Adams
Vice President, Marketing
Weidner Communications Corporation
40 Skokie Blvd.
Suite 300
Northbrook, IL 60062

Mr. Ronald J. Albaugh
Senior Associate Editor
Chemical Abstracts Service
P.O. Box 3012
Columbus, OH 43210

Mr. Ronald Aqua
United States-Japan Foundation
560 Lexington Avenue
New York, NY 10022

Mr. Frank Ballance
Action for World Development
2016 15th Street, N.W.
Washington, D.C. 20009

Mr. Herman Baron
Box 194
Moylan, PA 19065

Ms. Yoriko Beach
817 8th Street, #102
Laurel, MD 20707

Professor R. Byron Bird
Chemical Engineering Department
University of Wisconsin-Madison
1415 Johnson Drive
Madison, WI 53706

Mr. Michael Black
Account Data Group
2139 Wisconsin Avenue, N.W.
Suite #202
Washington, D.C. 20007

Mr. Justin L. Bloom
President
Technology International Inc.
11600 Georgetowne Court
Potomac, MD 20854

Mr. Edward L. Brady
Associate Director for
International Affairs
Administration Bldg., #A505
National Bureau of Standards
Gaithersburg, MD 20899

Mr. Cyril Braegelmann
Foreign Broadcast Information Service
P.O. Box 2604
Washington, D.C. 20013

Mr. Larry D. Brouhard
President
Intercultural Communications Service
2-1-14 Tabata, Kita-ku
Tokyo, Japan

Dr. Paul Brown
Director, Advanced Studies Program
Center for Advanced Engineering Study
MIT
Cambridge, MA 02139

Mr. John Clark
Center for Applied Linguistics
3520 Prospect Street, N.W.
Washington, D.C. 20007

Dr. Joseph E. Clark
Deputy Director
National Technical Information Service
5285 Port Royal Road
Springfield, VA 22161

Mr. Robert E. Cole
Fellow
The Wilson Center
Smithsonian Institution Bldg.
Washington, D.C. 20560

Dr. Samuel Coleman
Associate Director for Research
North Carolina Japan Center
North Carolina State University
P.O. Box 8112
Raleigh, NC 27695-8112

Mr. Alex R. Conn
STATS
No. 44 Thornton Avenue
Venice, CA 90291

Mr. Art Corte
OES/SAT Room 4333
U.S. Department of State
Washington, D.C. 20520

Professor Edward E. Daub
General Engineering
University of Wisconsin-Madison
Madison, WI 53706

Dr. Arthur Diness
Office of Naval Research
800 North Quincy Street
Arlington, VA 22217

Mr. J. David Edwards
Executive Director
Joint National Committee for Languages
3520 Prospect Street, N.W.
Washington, D.C. 20007

Mrs. Helen Eng
3111 Aberfoyle Place, N.W.
Washington, D.C. 20015

Ms. Deborah Erickson
Tokyo Electric Power Company, Inc.
1901 L Street, N.W.
Suite #720
Washington, D.C. 20036

Dr. Ellen S. Frost
Director of Government Programs
Westinghouse Electric Corporation
1801 K Street, N.W.
Washington, D.C. 20006

Mr. Hideo Fujinuma
Project Development Dept.
Nissho Iwai American Corporation
1211 Avenue of the Americas
New York, NY 10036

Mr. Nevil Garrett
President
Weidner Communications Corporation
40 Skokie Blvd.
Suite 300
Northbrook, IL 60062

Dr. Prosser Gifford
Deputy Director
The Wilson Center
Smithsonian Institution Bldg.
Washington, D.C. 20560

Ms. Susan Goldman
Director, Japanese Technologies
Translation Project
Courant Institute of Mathematical
Sciences
New York University
251 Mercer Street
New York, NY 10012

Dr. Lawrence Grayson
Department of Education
400 Maryland Avenue, S.W.
Washington, D.C. 20546

Ms. Marie T. Hall
7712 Tobruk Court
Hanover, MD 21076

Ms. Fumiko Mori Halloran
2939 Van Ness Street, N.W.
Washington, D.C. 20008

Mr. Tatsuya Hayashi
Manager, Software Laboratory
Fujitsu Laboratories Ltd.
1015 Kamikodanaka, Nakahara-ku
Kawasaki, 211 Japan

Ms. Pat Hill Hubbard
Vice President
Engineering & Technical Education
American Electronics Association
P.O. Box 1045
Palo Alto, CA 94303

Dr. Randall L. Jones
Humanities Research Center
3060 JKHE
Brigham Young University
Provo, UT 84602

Professor Eleanor H. Jorden
321 Morrill Hall
Cornell University
Ithaca, NY 14853

Mr. Toshihiko Kanda
Manager, Marketing Section
Japan Information Center of
Science and Technology
5-2 Nagata-cho, 2-Chome
Chiyoda-ku, Tokyo, Japan

Ms. Hiroko Kataoka
North Carolina Japan Center
North Carolina State University
Box 8112
Raleigh, NC 27695-8112

Mr. Yoshihiro Kawaguchi
Japan Information Center of
Science and Technology
5-2 Nagata-cho, 2-chome
Chiyoda-ku, Tokyo, Japan

Mr. Masaru Kawahara
Bravice International Inc.
Sumitomo Ichigaya Bldg.
1-1 Ichigaya, Honmura-cho
Shinjuku-ku, Tokyo, Japan

Mr. Wayne Kiyosaki
Foreign Broadcast Information Service
P.O. Box 2604
Washington, D.C. 20013

Mr. A. E. Klauser
Senior Vice President
Mitsui & Company (U.S.A.), Inc.
1701 Pennsylvania Avenue, N.W. #400
Washington, D.C. 20006

Mr. Tadashi Koretsuné
President
National Science and Technology
Corporation
5-17, 3-Chome
Kita-Aoyama, Minato-ku
Tokyo, Japan

Ms. Claire J. Kramsch
Department of Foreign Languages
and Literatures
14N-226
MIT
Cambridge, MA 02139

Professor Richard D. Lambert
Box 127A
Barnegat Light, NJ 08006

Mr. C. M. Lee
Code 432F
Office of Naval Research
800 North Quincy Street
Arlington, VA 22217

Mr. Richard B. Levine
Deputy Director
International Economic Affairs
National Security Council
Old Executive Office Bldg. #365
Washington, D.C. 20506

Mr. Hisao Matsumoto
Japanese Section, Asian Division
The Library of Congress
Washington, D.C. 20540

Mr. Masayoshi Matsumura
The Japan Foundation
New York Office
342 Madison Avenue, #1702
New York, NY 10173

Ms. Roberta H. Merchant
5420 Stormdrift
Columbia, MD 21045

Ms. Nancy Miller
Science Policy Division
Congressional Research Service
Library of Congress
Washington, D.C. 20540

Dr. David Mills
Dept. of East Asian Languages
and Literatures
University of Pittsburgh
1501 Cathedral of Learning
Pittsburgh, PA 15260

Mr. Carlos Montalvo, Jr.
Xerox Corporation
1616 Ft. Meyer Drive
Rosslyn, VA 22209

Ms. Masako O. Moore
1168 Kettle Pond Lane
Great Falls, VA 22066

Dr. Ronald A. Morse
Secretary, Asia Program
The Wilson Center
Smithsonian Institution Bldg.
Washington, D.C. 20560

Dr. David T. Mowry
Associate Director
Center for the Utilization of
  Federal Technology
National Technical Information Service
5285 Port Royal Road
Springfield, VA 22161

Ms. Janet Murray
Director
Athena/Annenberg Project
20B-231
MIT
Cambridge, MA 02139

Mr. Akira Nagasawa
Bravice International Inc.
Sumitomo Ichigaya Bldg.
1-1 Ichigaya, Honmura-cho
Shinjuku-ku, Tokyo, Japan

Mr. Myron C. Nagurney
Manager, Translation Program
National Aeronautics and Space
  Administration (Code NIT-4)
Washington, D.C. 20546

Mr. Kazuo Nakazato
Research Marketing Services
Japan Economic Daily
Kyodo News International
50 Rockefeller Plaza, #832
Associated Press Bldg.
New York, NY 10020

Mr. Tim O'Brien
Vice President, Marketing
Account Data Group
2139 Wisconsin Avenue, N.W. #202
Washington, D.C. 20007

Ms. Shizuka Ohkawa
Staff Assistant
International Programs
Westinghouse Electric Corporation
Waste Technology Services Division
Waltz Mill Site
P.O. Box 286
Madison, PA 15663-0286

Mr. Yoshio Osada
Mitsubishi Research Institute, Inc.
Metropolitan Square
655 15th Street, N.W., #880
Washington, D.C. 20005

Ms. Barbara J. Peterson
Manager, Information Service
3M Center 201-25-09
St. Paul, MN 55144

Mr. Andrew H. Pettifor
Office of Science and Technology Policy
New Executive Office Bldg., Room 5005
Washington, D.C. 20506

Mr. Roger Picciotto
Associate Manager, International
  Technology Coordination
The Procter & Gamble Company
Winton Hill Technical Center
6060 Center Hill Road
Cincinnati, OH 45224

Mr. Kevin P. Phillips
President
American Political Research Corporation
4312 Montgomery Avenue
Bethesda, MD 20814

Mr. Lewis R. Podolske
Technical Assistance Specialist
Office of Trade Adjustment Asst.
International Trade Administration
U.S. Department of Commerce
Washington, D.C. 20230

Ms. Carol Pompliano
House Committee on Science & Technology
U.S. House of Representatives
Washington, D.C. 20515

Mr. Joseph Price
Head, Science & Technology Division
Library of Congress
Washington, D.C. 20840

Dr. John Ratliff
Acting Dean, Language Studies
U.S. Department of State
(M/FSI)
Washington, D.C. 20520

Mr. Barton Reppert
The Associated Press
2021 K Street, N.W.
Washington, D.C. 20006

Dr. John P. Riganati
Technology 225-A219
National Bureau of Standards
Gaithersburg, MD 20899

Mr. Thomas Rimer
Chief, Asian Division
Library of Congress
Washington, D.C. 20840

Mr. Eugene F. Rowan
Divisional Vice President & Manager
of Federal Government Relations
J.C. Penny Company, Inc.
1156 15th Street, N.W.
Washington, D.C. 20005

Mr. Thomas R. Ruddy
Washington Representative
IBI Inc.
1735 Eye Street, N.W. #813
Washington, D.C. 20006

Dr. Robert A. Russell
Assistant Professor
Department of Asian and Near
Eastern Languages
Brigham Young University
4062 JKHB
Provo, UT 84602

Dr. William Salmon
Special Assistant to the Under
Secretary of State for Science,
Technology and Security Affairs
U.S. Department of State
Washington, D.C. 20520

Dr. Richard J. Samuels
Director, MIT-Japan Science
and Technology Program
MIT
Room E53-447
Cambridge, MA 02139

Mr. Tomoyuki Satoh
Box 194
Moylan, PA 19065

Mr. Senichi Shamoto
Senior Vice President
Nissho Iwai American Corporation
1825 K Street, N.W. #1103
Washington, D.C. 20006

Mr. David Shonyo
Director
Office of International Affairs
National Technical Information Service
Springfield, VA 22161

Professor Eugene Skolnikoff
Director
Center for International Studies
MIT
Cambridge, MA 02139

Ms. Pam Smith
OES/SCT, Room 4330
U.S. Department of State
Washington, D.C. 20520

Dr. Richard M. Spriggs
Senior Staff Officer
National Research Council
National Academy of Sciences
2101 Constitution Avenue, N.W.
Washington, D.C. 20436

Ms. Kiyo Stevenson
12009 Turf Lane
Reston, VA 22091

Mr. Toshitomo Tajima
Bravice International, Inc.
Sumitomo Ichigaya Bldg.
1-1 Ichigaya, Honmura-cho
Shinjuku-ku, Tokyo, Japan

Mr. Phillip Thibideau
Manager, International Scientific
and Technical Activities
National Aeronautics and Space
Administration
Headquarters (Code NIT-4)
Washington, D.C. 20546

Mr. Richard Thompson
Deputy Director, Center for
International Education
U.S. Department of Education
400 Maryland Avenue, S.W.
Washington, D.C. 20546

Mr. James M. Unger
Department of East Asian Languages
and Literature
University of Hawaii-Manoa
1890 East-West Road
Honolulu, HI 96822

Mr. Shigeko Uppulri
Information Division EMCTIP
Oak Ridge National Laboratory
Area Y-12, Room 9224
Oak Ridge, TN 37831

Mr. Galal Walker
Assistant Professor
Department of East Asian
Languages and Literature
Ohio State University
276 Cunz Hall
Columbus, OH 43210

Dr. Charles Wallace
Division of International Programs
National Science Foundation
1800 G Street, N.W.
Washington, D.C. 20550

Ms. Patricia E. Winton
Staff Assistant
Nuclear Center - MNC561
Westinghouse Electric Corporation
Northern Pike/Haymaker Road
Monroeville, PA 15146

Ms. Dalena Wright
Office of Congressman Mineta
U.S. House of Representatives
2350 Rayburn House Office Bldg.
Washington, D.C. 20515

Mr. George Wright
Office of Naval Research
800 North Quincy Street
Arlington, VA 22217

Dr. Raymond H. Wylie
Department of International Relations
Lehigh University
208 Maginnes Hall 9
Bethlehem, PA 18015

Mr. Takehiko Yamamoto
Bravice International Inc.
Sumitomo Ichigaya Bldg.
1-1 Honmura-cho, Ichigaya
Shinjuku-ku, Tokyo, Japan